"Understanding New View Safety is one thing: enacting it in everyday leadership is another. Through Empowered Solutioning and the Integral Model, David Wollage provides a practical bridge between the two...combining coaching, safety expertise and systems thinking to help leaders move from knowing differently to leading differently."

Kym Bancroft, *Director, New View Safety, Australia*

"This is not a soft approach to safety. It is a harder one, as it holds leaders accountable for the conditions they create, not merely the rules others break ... Curiosity is not the opposite of accountability. In my experience, it is the only reliable route to it ... Nothing in this book lets a leader off the hook. It simply changes the questions they are required to answer."

Martyn Campbell, *Former WHS Regulator and Chief Executive, SafeWork South Australia, Australia*

"Everything starts with leader knowledge, language, and behaviours, and David captures all three. This could be a personal how-to guide for improving you and your team's leadership and coaching"

Rob Fisher, *President and Director of Operations, Fisher Improvement Technologies, USA*

"What I appreciate about David's book is how he reinforces that there isn't one right approach for every situation. He nudges you to look beyond the surface to personal and societal conditions, and to get real about your own values before trying to lead others. Throughout, his reflective questions and coaching tone make the ideas practical and invite you to be curious about your own leadership style and how that might influence a psychologically safe workplace. I believe there's no greater leadership superpower than curiosity..."

Cheryl Pocklington, *Director of Regulatory Modernization, Compliance and Enforcement Secretariat, Canada*

I0875884

New View Safety Coach

Workplace safety and leadership are facing evolving challenges, and this book explores the intersection of two powerful global trends: executive/organisational coaching and New View Safety methodologies. By defining these trends and demonstrating their collaborative relationship, it presents a compelling case for how they can be combined to drive meaningful change in individual leaders and organisational culture to create safer workplaces.

New View Safety Coach: Empowered Solutioning is a quick and neat guide for professionals in understanding how to introduce safety methodologies in a working environment. Evidence-based strategies show how coaching can support the implementation of New View Safety, leading to improved leadership practices and enhancing safety outcomes. The book offers practical guidance on integrating these methodologies, empowering leaders to initiate, sustain, and evolve safety initiatives effectively. The reader will learn how to stay ahead of regulatory demands and create a more resilient organisational culture wherever their workplace may be.

This book is an ideal read for health and safety professionals, executive leaders seeking innovative ways to meet rising safety standards, business professionals looking to shift toward a coaching role, and operational leaders eager to embed New View Safety principles in their organisations to deliver better outcomes.

New View Safety Coach

Empowered Solutioning

David Wollage

CRC Press
Taylor & Francis Group
Boca Raton London New York

CRC Press is an imprint of the
Taylor & Francis Group, an **informa** business

First edition published 2027
by CRC Press
2385 NW Executive Center Drive, Suite 320, Boca Raton FL 33431

and by CRC Press
4 Park Square, Milton Park, Abingdon, Oxon, OX14 4RN

CRC Press is an imprint of Taylor & Francis Group, LLC

ISBN: 9781032966021 (hbk)
ISBN: 9781032967196 (pbk)
ISBN: 9781003590538 (ebk)

DOI: 10.1201/9781003590538

Typeset in Times
by Deanta Global Publishing Services, Chennai, India

Dedicating a book to someone, or a group of people, should not be a casual affair. Books, unlike blog posts, social media comments or websites, cannot be retracted once they are out in the world. There is no erase button. No factory recall. The book stands as a permanent testament, not just of the author's thoughts and opinions, but also the value they place on the individuals to whom they have dedicated the book.

There are many people I could dedicate this, my first-ever book, to. Some of these people are trusted and respected mentors, close friends, or both. My hope is that, if they are reading this, they already know how much I value them.

How much I trust and respect them. How my world would be a darker place without them. That being said. This book is dedicated to two very important people to me. Roger Coombe and his truly amazing partner, Lesley Dawson. Roger is my stepfather, although this title doesn't do him any justice. We only got to spend a short time as father and son; however, many years later, we reconnected, and he has once again taken the responsibility of father figure to me. No questions. No doubts. Just love and acceptance. And for this Dad, I am and will always be eternally grateful. Leslie, you have accepted me, cared for me, and shown me love beyond what could ever be expected of you. For this, I thank you from the bottom of my heart.

I love you both.

Contents

About the Author

David Wollage is a safety professional and organisational coach based in Perth, Australia. He holds degrees in Safety Leadership from Griffith University, Business Psychology from the University of Western Australia, Australia, and Organisational Coaching and Leadership from Charles Sturt University, Australia. Certified as a COHS Practitioner and an Organisational Coach through the Institute of Executive Coaching & Leadership, Australia, Wollage integrates safety expertise with evidence-based coaching practice to support leadership and culture transformation. His YouTube channel, New View Safety Coach, dedicated to safety leadership and coaching, has grown to over 86,000 subscribers.

Why Write This Book?

I've never been particularly good with details, remembering dates, names or statistics, but for some reason I've always tried to see how concepts either are or could be connected on different levels. This hasn't always been easy for me, as school only required people to have recall capabilities, as opposed to being conceptually minded. My frustration would surface occasionally, and I mostly felt like a fish being asked to climb a tree! I didn't do badly at school, and what I did manage to achieve came only with lots of effort. As an example, instead of just accepting the use of "x" and "y" being used in maths equations, I would find myself asking questions like "Why those particular letters?" and "What relevance do they have to the equation?" I remember some of my friends could read information, not question any of the symbols or use of arbitrary letters, and then later access that information (visually) in their minds. An aspect of memory recall. This type of recall always eluded me, and I felt inferior because of it. It's only recently that I've discovered that there is such a thing called Aphantasia and that I (highly likely) have it. Aphantasia is a condition whereby the person cannot voluntarily create mental images in their mind's eye – nothing, no images of loved ones, favourite pets, places, diagrams, formulas, systems, or even a representation of self! When I close my eyes and think of people, places or objects, I see only blackness or those strange, floating blobs of colour. One of the consequences of this condition is that people with it tend to think more conceptually and abstractly than those who don't have it. Aphantasics can also have increased focus and attention due to the lack of mental visuals; this has definitely been a bonus for me as I sit with a coachee and create a space for them to place their meanings and interpretations of themselves, and the world as they see it, without any mental visuals interfering with my process of being the coach.

So how does this impact why I decided to write this book? I'm always naturally trying to push my understanding of topics up to as meta-level as I can, the least detailed but most connected view. My proclivity growing up has stayed with me. So now I seek out connections between topics. I look for commonalities across theories. The more connections I can make with other ideas, the more easily I can relate to and remember something.

I've been working in Safety for a while now, and I've been exposed to Organisational Coaching for much of that time. As soon as I was exposed to

Organisational Coaching, I immediately realised that there is a strong connection between these two skill sets and professions, safety and coaching. As I have been seeking more and more knowledge in relation to Organisational Coaching and New View Safety methodologies, the connections have become clearer and more evident. Delving into these sometimes subtle, sometimes blatant and overt connections between these two disciplines, it began to open up a new world for me, a world where Organisational Coaching and safety (more specifically New View Safety) share a largely symbiotic relationship, supporting each other with very similar values, purposes and principles! It's a world where wisdom through the times can be incorporated into our modern lives through the catalyst of these two practices, and we can benefit both individually and in our roles within organisations. When I saw this growing connection, I knew it was the place I wanted to work, and I knew I had to tell others so that they could benefit. To be very clear and upfront, I am not the first to realise that these two disciplines work so well together; there are other individuals in the world who have also made this connection and written articles on this very topic, which we will cover later. There just isn't a book written about this relationship – until now!

THE BOOK OF 2 PARTS

This book has a theme of two, and it is how the book is structured. As you will see, the book is structured into two parts:

Part I deals with very important foundational questions, such as:

- What is New View Safety?
- What is Organisational Coaching and how can individuals benefit from it?
- How are these two disciplines symbiotic in their approaches and desired outcomes?

It is framed for the individual. How New View Safety can (sometimes) be difficult for individuals to engage with, and why. It also covers how the use of Organisational Coaching can facilitate a way forward for individuals to gain a deeper understanding and appreciation of New View Safety principles.

Part II demonstrates how, when using a coaching leadership style, one person can influence, motivate, and inspire others around them, both for personal development and for adoption of New View Safety principles.

Introduction

I was cornered. I had to listen to what this person was saying, even though I was preparing for a meeting and had to get a report done by the close of business that day. I was busy, very busy. I knew from previous experiences that this person had a habit of complaining, and not just that, but also taking forever to do it! I no longer worked there; I was there now as an external consultant filling in while they found another safety leader.

As I listened, my internal voice was saying, "Not now, I really don't have time for this!" As I grappled with this internal dialogue, I heard the word that stopped me in my tracks – "Asbestos!" Now I was listening. I had picked up on some keywords from the five minutes or so of their talking; something about an air conditioning unit in their shared offices, I think! Was there asbestos in the air conditioning unit? How did they know? How long had they known? My mind went into overdrive with a million questions and the consequences of these people being exposed to asbestos via an air conditioning unit. Was the asbestos friable? What would happen to these people? What would happen to me?

I let them continue to talk for a while longer as I concentrated on trying to remain calm, and then I asked, "How do you know for sure there is asbestos coming from the air conditioning unit?" Their answers confused me. They didn't "know for certain"; they suspected, and they wanted it to be raised at the next Safety Representative meeting. This was their prerogative; they were a Safety Representative, and this was their role, to raise safety concerns to the organisation. I agreed with them, but I wanted to understand more about why they suspected asbestos first, so I asked them how long they had "suspected" that there was asbestos coming from the air conditioning. Again, their answer was vague, "oh, for a while maybe." This didn't make sense, and I was beginning to get annoyed. Did this person understand the severity of saying they suspected asbestos? Did they understand that once this was put into motion, there would be massive amounts of effort put into verifying the presence or absence of asbestos? And all of that on the hunch of someone who was giving me vague answers and who didn't seem to understand the consequences of what they were saying!

Two weeks prior, I had been put through my very first Organisational Coaching course. We had been taught the GROW model of coaching and how to structure conversations using it. GROW is an acronym that stands for:

Goal/s
Reality
Options
Will

An important part of the coaching conversation is to establish what the person you are talking to wants to achieve – what is their goal?

I took a breath and wondered how I should proceed. I knew my emotions were heightened, and I figured the person would soon begin to pick up on my frustration. For some reason, I remembered the coach training, the GROW model, so I asked, "What is it you want to achieve? What is your goal in all of this?" To my surprise, after giving it some thought and checking no one could hear, they said, "To be honest, I just want the temperature turned up as I really don't like the cooler temperature!" I was shocked. Did I hear correctly? This is a temperature issue, not an asbestos issue? I asked why they had mentioned asbestos if it was a temperature issue. They said, "Well, the air conditioning unit annoys me all day, every day! I was just trying to get rid of it!" I was shocked. This isn't a bad person, I thought. Annoying at times, maybe, but not bad. I pulled myself together and went back to the GROW model. OK, I said, what options do you have other than raising an asbestos concern for dealing with your dislike for the current temperature of the air conditioning in your office? They paused and thought.

Then they said, "Well, I guess I could speak to the other people in the office and see what they think?" And what else I said, "Or if everyone else likes the temperature, I could wear a thicker jumper?" I agreed with their options, reconfirmed there really wasn't a concern with asbestos, asked when they would do all this by (the "W" part of GROW), and then went off to do the tasks I had to do.

Although not a classic case for determining someone's goals, what the above story does demonstrate is that sometimes all we need is someone to ask us what we really want – and I mean what we "really" want. Part of coaching is to gain clarity on what the person's goal is, because sometimes they don't know what their goal is. And sometimes they mask their real goal with something else! Towards raising an alarm about asbestos, for example!

It wasn't until the next day, when I was in a morning huddle where I worked, that I raised the story. I told everyone how an asbestos concern had suddenly been downgraded to a temperature issue, all by asking what someone "really" wanted. Others at the company had been trained in Organisational Coaching longer than I, and they weren't as surprised as I was.

I needed to understand more about this Organisational Coaching and how it could benefit safety.

PART I

New View Safety in One Page (Maybe Not!)

1

New View Safety represents a significant shift in how we approach workplace safety. It challenges the traditional or "old view" of safety, which can be described as "focusing almost entirely on preventing accidents, incidents and injuries by stopping things from going wrong" or "where the number of things that go wrong is acceptably small." In other words, where safety is measured by its absence, by the number of things that go wrong! This "old view" approach has created systems and approaches to enable it. Still, it is these very systems and processes that have come under scrutiny as not being fit for purpose in 21st-century organisations. This can be evidenced by a plateauing of safety across multiple industries; a state where the decreasing number of fatalities is either stalling, or worse still, actually reversing and beginning to rise. Why could this be? Have we stopped caring as much about safety? Is safety not as important as it once was? Are our leaders slowly getting worse at their job? I know I'm not alone when I say that this is not the case.

WHY IS IT HAPPENING THEN?

When we analyse old-view safety, we see that there are some inevitable outcomes and thought processes that we still hold on to most dearly. They have formed part of society's presuppositions in relation to safety, and for good reason, they have been around for a long time! Let's take a look at them so that we can begin to articulate exactly how history (and maybe even evolution!) has shaped how we have "done" safety.

DOI: 10.1201/9781003590538-2

As I've already outlined, the Old View of safety predominantly focuses on failures, accidents, rules and errors in those rules. This had many positive attributes as the Industrial Revolution began to dawn, and industries began to industrialise manufacturing, transportation, and other heavy industries. However, this "perspective" on safety has a limited scope for improvement. To steal a line from Todd Conklin – "imagine trying to figure out how to have a great marriage by only asking divorcees for advice!" Your data set will be focused purely on what could go wrong, what actually went wrong, and the inevitable consequences of that. As I said, this is very useful, to a degree, but it only gets us part of the way in modern times.

If we know this to be true, our fascination with focusing on error and mistakes, and the limitations it has, why do we find it so easy to do? Why have we held onto noticing "error" and "mistakes" so much? I've often thought about this, especially when I find myself reacting to news of an incident in this exact way. I'm not an anthropologist or clinical psychologist, so this is out of my realm of competency; however, I have a theory I thought I'd share. Evolution is to blame (David, can you stop blaming, please! – *Editor*). If, over the course of our evolution as a species, we weren't noticing when things went "wrong," could we have made it as a species? Would we have been able to learn from all the things in the world that could have killed us? Dark caves with growling inside are best avoided. Witnessing another human eating a certain plant and then dying from it. With the amount of danger we faced (until relatively recently in historical and anthropological terms – again, I'm no expert!), it may be a feasible argument as to why we as humans have such a strong proclivity for noticing potential for error and mistakes. After all, there is only a limited way things can go well, and even when they do go well, it didn't pay to celebrate too much and brag about it, for fear of others taking retribution out of jealousy. But outcomes can vary greatly on the negative side, from an inconvenience, all the way to death! I believe this is one of the main reasons we are so attuned to notice when things don't go as desired. Could it also begin to explain complacency when things usually go as desired? On a side note, we will revisit this aspect later when we discuss biases and how they influence the very way in which we "see" and interact with the world.

This myopic focusing on error and failure has other consequences on how we enact safety; not least how, when we focus on error and failure, we set ourselves up emotionally, which generally leads to an emotional "reaction" as opposed to an emotionally stable "response." The emotional state we adopt will directly and indirectly influence how safety is enacted by ourselves and the organisation. Don't believe me? Here's an example of how what we focus on affects our emotional state:

Scenario

A forklift hits a bollard during a rush job. No one was injured, but it was close.

1. Focusing on the error and lack of compliance (fear, annoyance, avoidance) Reactive: "You were careless and put people at risk! This is unacceptable. Why did you do it?"
2. Noticing your initial reaction but being able to moderate your emotions (control, curiosity, learning mindset)

Responding: (I'm noticing I'm wound up and frustrated. Breathe.) "What pressures and design choices made this the *easier* move? If I understand those, I can make the safer way the *easy* way for everyone."

When we focus only on the negatives, we react negatively. When we react negatively, not by choice, I may add, but because it's how we view "safety," we indirectly tell those we are responsible for that we think they are bad people who don't care enough! They are stupid and need to fully appreciate the outcome of their actions on those around them more clearly! This leads to obvious control measures such as more training (for the guy who's been driving a forklift for 25 years!), refresher courses, public shaming – and yes, even dismissal. When we see safety as purely preventing bad things from going wrong, we see the solutions as "fixes" to the undesired behaviour, logically. But when we reframe how we view safety, we directly impact how we "respond" to safety concerns. Consider the example given. Which approach is more likely to learn about what "actually" caused the incident and prevent it in the future?

Having a world view of safety as being "the absence of negative events" and how we emotionally react to incidents isn't the only downside of Old View Safety. Here are a few more examples.

SAFETY INDICATORS

This topic has already been covered extensively by people far more qualified than I. So, in the spirit of defaulting to those others, I will keep this part brief. Another conclusion of only focusing on failures and accidents leads us to measure those unwanted events, which is perfectly logical and useful to a degree. The metrics we use, commonly Lost Time Injury Frequency Rate (LTIFR) or some other measurement scale, focus us on the attainment of an arbitrary number in the belief that if we can remove enough injuries, incidents, and violations, we will have a safer system of work. Before your thought process leads you to believe I am against LTIs or other methods of recording lagging indicators, please relax! I'm not for the abolishment of lagging indicators entirely. I do, however, view such approaches through a psychological lens and appreciate, through others and my own experiences in safety, how such metrics can be misused, misappropriated, and can serve as a false goal, especially if improving actual "safety" is your desired outcome. This again comes back to that defining question we should all ask ourselves – What do I mean when I say "safety?" (don't worry, there is a"Questions to poke your brain" section at the end of this, and all chapters, to have you reflect on things like this). I go into lagging indicators later in this chapter, so if this topic is particularly contentious for you, there isn't long to wait!

NEGLECT OF NOTICING SUCCESS AND LIMITED LEARNING OPPORTUNITIES

An almost unavoidable consequence of predominantly focusing on what goes wrong is the creeping lack of attention to what goes "right." This translates into a lack of acknowledgement for when things go "right," even during those adverse and dire conditions, and this leads to a stagnation in learning about the safety of work. Success, therefore, is usually put down to the existing systems and rules in place, as opposed to the adaptations and creativity of the people at the front line, who, I promise you, didn't follow the rules or procedures to the letter, despite what you may think! A cornerstone of New View Safety is the desire to learn from these successes, as well as errors, as it is these successes that give us a larger dataset of information from which to continually improve, especially if the way something "worked" wasn't the way it was written in the work instruction!

If we mix in a bit of curiosity when we recognise these successes, we end up with questions such as:

- *What aspect of the process was critical to the success? (great for bow-tie development).*
- *Did it go exactly as documented, or do we need to update the documents so others can learn from this success?*
- *Are there any other aspects of our work that could benefit from what worked well?*
- *It's great that we succeeded, but what might have ruined it for us?*

Depending on the complexity of your organisation, and therefore the interconnectedness of it across more systems internally and externally, this continuous learning becomes paramount to success (read "safety").

Now, before I go any further, I want to make something crystal clear. New View Safety, as I understand it, is not advocating for a complete 180-degree change to all safety systems within your organisation! It recognises and highly values the strengths that the Old View brings; New View wants to "add to" the Old View and enhance the overall safety management of an organisation by complementing it and addressing the areas that are now needed in modern-day organisations such as the growing complexity of systems, increased interrelated business structures between contractors and other stakeholders, and increased dependency on human capability within these complex systems. You could say New View Safety isn't a standalone approach or methodology; it is another piece that addresses modern problems not already addressed by the Old View. A partnership, if you will. Let's hope this helps with the current safety tribalism on platforms such as LinkedIn, where people, again logically, feel the desperate need to defend their presuppositions and beliefs around which safety methodology is the best, and why! I get it. When people (including myself) believe their perspectives and beliefs are being questioned, the counter we have in our brain, which keeps track of our social status, is activated, and we become defensive. I believe it was Clive Lloyd who used the term "Safety Tribalism" in a LinkedIn post, and he would know.

But hang on, if it's not an entirely "new" approach to safety, why is it called New View Safety? Good question. The "New View" of safety is not entirely new, but rather a collection of concepts and methodologies that have evolved from, or built upon, several decades of prior research and existing practices, while also introducing significant shifts in perspective. It's these presuppositions that are paramount, and we will be addressing them in detail as you work your way through the book.

> Since we cannot change reality, let us change the eyes which see reality.
>
> Nikos Kazantzakis

The actual origins of New View Safety come from Cognitive Systems Engineering (CSE), which was developed in the 1970s onwards by pioneers such as Rasmussen, Hollnagel, and Woods. CSE focused on human error in safety-critical systems and drew on disciplines such as cognitive psychology, engineering, and systems thinking. The term "New View" was actually only coined by Dekker in the 2000s, who translated the output from CSE and compared it to an "old view" of safety. Over time, the "New View" has come to mean various approaches such as Resilience Engineering (RE), Safety-II,

Safety Differently, and Human and Organisational Performance (HOP). These concepts are often grouped together as "New View Approaches" due to their shared foundational concepts, such as defining health and safety as an emergent system property, rather than the absence of injuries.

So then, what "exactly" is New View Safety I hear you say??

Think of New View Safety this way. It's an evolution of safety practice which utilises many social and engineering sciences to account for the ever-increasing complexity of the world in which we live and work.

That was fairly succinct for me, considering I'm known to waffle on at times!

NEW VIEW SAFETY IN ONE PAGE (MAYBE NOT!)

So now we have outlined what New View isn't, i.e., Old View, and we have laid out its place within modern organisations, let's take a closer look at what exactly New View Safety is.

A surface-level appraisal of New View would go something like this: New View, with its roots in psychology and systems engineering, recognises that human error is ALWAYS inevitable and blaming humans for being human is counterproductive at best, and destructive at worst.

New View Safety looks to move the focus from blaming the individual to focusing on the systems the individuals are using. If you must "blame" something, blame the systems! By proactively learning about what goes well within the organisation, New View seeks to do more things well, as the more we do that works well and goes to plan, by default, the safer we will be. In essence, New View Safety emphasises understanding how work is successfully accomplished in complex systems, recognising human adaptability as a strength, and promoting a culture of learning from both successes and failures, to enhance overall system resilience.

As with most topics worth discussing, they are generally complex and multifaceted in nature. Although we generally like ideas, concepts, and beliefs to be confined within parameters, it isn't always useful, as it sparks more debate than it clears up. That being said, for those of you who are new to the concept of New View Safety, I will endeavour to put New View Safety in simplistic boxes and compare it to Old View Safety.

OLD VIEW SAFETY	NEW VIEW SAFETY
Workers are generally the problem	Workers have more solutions to offer
Focuses more on behaviour	Focuses more on systems and psychology
Learning is generally inhibited	Learning is rewarded and embraced
Fewer incidents mean more safety	Safety is not defined by the absence of incidents, but by the presence of capacity
Good employees behave as we want them to	Context and systems drive behaviour
Errors are bad decisions made by incompetent people	Error is an intrinsic part of work (and life!), and no one is error-free

(Scan the QR and search my YouTube videos for "What are HOP and Learning Teams")

Let's take a closer look at some of the most contentious aspects of New View Safety that seem to generate the most debate and conflict.

ERROR AS SYMPTOM; SAFETY AS SYSTEM

How upset or annoyed do you get when you're brought the wrong food at a restaurant? What about when someone gives you the wrong drink at the bar, someone nearly sideswipes you at an intersection where you have right of way, or, and here's the big one, how do you feel (be honest now) when one of your workers (insert most relevant option):

- Damages an expensive piece of equipment.
- Operates outside the scope of a procedure.
- Hurts themselves or others.
- Misses an obvious detail and brings an entire system or project to a standstill.

Likelihood, unless you are a saint and pure of heart, one of the above examples will have resonated with you, and you can recall the negative emotions felt at the time. This is a PG-rated book, so I'll keep expletives out of it, but the following won't be far from the mark:

- What an idiot!
- Are you stupid?
- How do you get by in life?

And so on.......

Now, let's reverse the focus and put ourselves directly in the line of "blame" in the following examples:

- You can't remember where you parked in a multi-storey car park.
- You lose your car keys again!
- You get into your new car and activate the wipers instead of the indicators.

Now this is where it gets interesting. When processing our own "errors," we are much more likely to factor in external events and mitigating circumstances, rather than immediately condemning ourselves as one of the above expletive descriptors.

- You can't remember where you parked in a multi-storey car park. "Well, it's a big place with lots of cars and all the floors look the same!"
- You lose your car keys again. "Who moved them this time?"
- You get into your new car and activate the wipers instead of the indicators. "Damn European car manufacturers! Can't they just put it on the same side as everyone else!" (This works here in Australia; it may not be applicable to you if you live in Europe!)

This phenomenon of attributing blame to others' personality or character flaws (as with the first set of examples) in the event of an error, and not treating ourselves the same way, is called, helpfully enough, the Fundamental Attribution Error (FAE). FAE is a bias we all have to a greater or lesser degree, and we will cover it in more detail later in the book, along with Confirmation Bias, or Attentional Bias as it's sometimes called (in case you need to know before we get there, Confirmation Bias is when we believe we already know how the outcome came about, and we seek only that evidence to support our hypothesis). Or another tricky rascal who creeps up a lot in safety, Hindsight Bias, where we view the incident in reverse, from the end point backwards, and only see "our" logical decision path given what we already know about how the incident occurred. This negates the realities and complexities experienced by those involved.

> Men are not disturbed by things, but by the views which they take of things.
>
> Epictetus, *Enchiridion* 5

The reason I point out FAE is that it highlights the unfairness and harshness by which we judge others over ourselves at the time an "error" is made. And this, my friends, is one of the biggest reasons why the concept of "error being normal" garners so much attention, from both sides of the debate. It can be difficult for us to apply the same external factors and system variability we readily allow ourselves when it's others who make the error. Error is judged harshly, and the people involved, more harshly!

But wait, David, I hear you cry; when I lose my car keys or forget where I parked my car, no one dies, unlike with my workers at work! So the consequences are not comparable!

True, the consequences may not be comparable. Let me frame it like this for you, then. Statistically, the most dangerous activity humans engage in daily is driving. An average of 3,200 people die every single day from motor vehicle-related incidents on roads. I'm going to make a general sweeping statement here. Most people reading this book will either drive a motor vehicle or be a passenger in one, most every day.

So, in the interest of fairness, let's reverse the tables and look at ourselves for a moment.

Likely every single day we engage in the MOST dangerous activity that kills the most humans out of all other activities we do. Now, I want you to put this book down and stop reading it if you can honestly say you have never made an error/mistake while driving! Go on, I'll wait!

If it's the most dangerous activity we as humans can do daily, why do you still make mistakes and errors? Don't you care enough? Do you need more training?

Probably none of these. It's just we are humans, and as the saying goes, "To err is human, to forgive divine."

Error is a constant feature, not a bug! Entropy comes to mind here. The sooner we become comfortable with this "bug," the sooner we can make peace with it, and learn from it.

Complacency is an interesting concept. Dr Todd Conklin says it's an inevitable outcome of a stable system, and I tend to agree. But here is another viewpoint I want to put forward. If we look at complacency as "self-satisfaction with yourself and your abilities despite the presence of hazards," a fairly reasonable definition in relation to safety, and we look at the exact opposite, being in a constant state of alertness. Who among us can honestly say they can achieve a constant state of alertness? And even if you could, it would come with its own negatives, such as fatigue, being easily startled, difficulty focusing, and overreacting to situations. The truth is, when we critique others, we don't just attribute errors to their personality or their potential character flaws (FAE), we also expect perfection from them at all times, Other-Oriented Perfectionism (O-OP). This brutal combination of FAE and unreal expectations (O-OP) is what drives the constant debate between factions when discussing the New View Safety concept that error is normal.

New View Safety treats human error as a signal that system conditions, i.e., tools, processes, workload, incentives, environment, etc., made the behaviour make sense at the time.

What to do about it then?

As mentioned earlier in the book, we will address dealing with biases and axioms later. But for now, let's look at an alternative approach to FAE and O-OP. If we adopt the viewpoint that error is normal (if you're not here yet, just go along with me for a moment), it reduces the fear that often accompanies incidents; we expect errors in complex work (or driving!) and can focus on learning rather than blame. It also frames our conversations (investigations) with curiosity and an empathetic approach when we realise that we ALL make mistakes and errors. Once we get to this place of curiosity and empathy, we can then begin to dig deeper than surface-level compliance. We can begin to focus on the "systems of work," rather than the "workers of systems." It is then that we can begin to really understand the "how" of the error, and not just the "why" or (more commonly) the"who!" In short, we can begin to appreciate fully that safety is an emergent property of interactions between our people and the systems they use; and we can focus our attention accordingly to find real controls, and not just (indirectly) saying "our systems are perfect, but our people aren't!"

How much do the systems and environment influence our behaviours?

When discussing how systems and context influence our behaviours, I usually reference one or both of the following examples:

1. Supermarkets all over the world know to place their most bought items (milk, bread etc.) at the furthest point from the store entrance. Why? Because then you have to walk past all their other "offers," and you are more likely to purchase more than you came for initially. They also use certain lighting that reduces blink rates (open eyes see more, ergo we buy more); they also use certain aromas (baked bread, pizza, coffee etc.) to make us feel hungry (hungry people buy more food!).
2. Traffic engineers and traffic controllers all utilise "systems" to influence your driving behaviour. Traffic engineers will decrease the length of, and the gap between, road markings before a known hazard (sharp bend in the road, changed road conditions etc.) as this decreased gap gives the driver the illusion they are driving faster than expected and subconsciously prompts them to (A) slow down, (B) become more aware of their surroundings. Traffic controllers utilise a more mobile version. When they need to slow traffic past a worksite close to the road, they will decrease the distance between cone placements and also slightly narrow the available road width in order to increase awareness and decrease speed. All of these are systems that moderate behaviour towards a desirable output.

Those are PG-rated examples that focus on subtle behavioural changes, but having recently read a book by Browning (1992) titled *Ordinary Men: Reserve Police Battalion 101 and the Final Solution in Poland,* I have a more extreme example I'd like to share with you.

Browning outlines a detailed historical study of a specific group of the German Reserve Police Battalion 101. These weren't highly trained military men; in fact, most of them had little to no military training and were considered too old for military training anyway. They were "ordinary men" from Hamburg. Bakers, butchers, carpenters, shop clerks. People just like us. And yet this group of "ordinary" men, serving as a low-level military police unit, became known as one of the most violent and murderous units at the time. How can this be? How can ordinary people commit extraordinarily horrendous acts? Brown's central argument is that, due to social and psychological pressures such as peer pressure, obedience to rules, situational expectations, and gradual desensitisation, these men (although not all, it must be said) gradually became mass murderers! An extraordinary example, I think you'll agree, of how our behaviours can be contextually manipulated through the environment and systems we are exposed to and a part of.

Less violent examples also exist, such as the Stanford Prison Experiment (Philip Zimbardo) and the very famous Milgram Obedience Experiment (Stanley Milgram). There is also some evidence to suggest that elevated disease and mortality threats can create psychological conditions that may make conservative or authoritarian political stances more appealing for some people, especially where cultural alignment allows. But I'll let you do your own research on this one, as it appears to be quite contentious!

None of the above-mentioned behaviours would have been possible if the participants had merely received a document telling them to do a particular thing. The systems and context (environment, expectations, human need for security and acceptance, etc.) all contributed to shaping their behaviour, as it does with yours when you're driving.

Now, of course, these examples show that people's behaviour can be "manipulated" to achieve unwanted at best, and plain horrendous at worst, outcomes. Conversely, the same can be said for achieving good and productive outcomes. Positive Psychology is a classic example. If we acknowledge that "safety" is an emergent property of all the complex social and systems interactions our workers deal with every day, then we can begin to see how we can begin to shape the desired behaviours we want, other than just mandating it with a safety document, or using blame and fear! As we've seen (and no doubt know all too well ourselves!), documents alone aren't enough.

VARIABILITY AS RESOURCE

One can define New View Safety as an evolution of safety practice which utilises many social and engineering sciences to account for an ever-increasing complexity of the world in which we live and work.

And unlike Old View Safety, New View Safety recognises that within these complex socio-technical systems, there will always be complex interactions that require performance variability from those doing the work. This links back to the previously mentioned fact that systems in the early 1900s were not as interconnected or complex as they are now. (Even driving is a complex operation if you really think about all the variabilities that can and do happen, constantly! Could you drive to work without adjusting your performance to match the environment or current context?)

(Scan the QR and search my YouTube videos for "Saving lives at sea")

So, what does this "Performance Variation" actually mean? It means that performance variability or performance adjustments are a sine qua non (one cannot work without the other) for the productive and safe functioning of complex socio-technical systems unless, as just discussed, those systems are extremely simple. In Old View Safety, this "variability" of human performance is seen as a liability and a threat, to be addressed by (yes, you guessed it!) more rules, regulations, training, and punishment. New View Safety recognises that these complex systems work exactly because people are able to adapt, identify, and overcome the myriad of complexities these socio-technical systems throw their way. (If you don't know what variabilities your organisation faces, just ask your frontline workers and middle management; they are already fully aware, and your best resource.)

Ashby's Law of Requisite Variety states that "only variety can absorb variety." I recently wrote a blog about this very topic. Below is a quote from my blog

> In simpler terms, to effectively respond to the diverse and unpredictable challenges within an environment, a system must possess an equal variety of responses. This idea, although originating in cybernetics, has powerful implications for modern safety practices, particularly in the realms of New View Safety.

(Scan the QR and search my website blogs for "Ashby")

Leaders therefore cannot simply demand resilience or better safety performance; they must actively ensure their teams have the necessary resources and support to successfully address the inevitable variability over the long term, which includes fostering performance variation that meets both safety and production needs. And it is these variations to successful work that New View Safety proposes we learn more about, since successes are far more abundant than failures in most systems (e.g., a 1:10,000 ratio of failure to success). Focusing on the 9,999 times things go right offers countless opportunities for learning, as opposed to the relatively rare times things go wrong. If our focus is on improvement, as opposed to only stopping bad things from happening, we will build greater resilience, better efficiency, and in the process, we will gain engagement and trust from the people doing the work.

I know it's easy to ignore things when they are working as intended. We don't notice the walls of our house until they show signs of cracking. We don't really notice our computers until they stop working. However, when there are parts of our lives we want to improve, like our fitness, education, or work-life balance, we focus on what worked for us, and then we look to either replicate it or improve on it the next time. These incremental improvements focused on what worked well and accumulated to lead us towards our desired state, whatever that is.

Let me put it this way, as I like to imagine conversations:

Safety Manager: "We've had several incidents whilst trenching. Let's get to the bottom of what caused them and make sure they don't happen again."

A fairly standard approach I think we'd all agree. Now, consider the following:

Safety Manager: "We've had several incidents whilst trenching, let's go and see what we can learn. Also, on the way back, let's drop in and see the cabling crew and ask what's working well, and if it can be improved."

There is nothing morally wrong with the first approach. It's a lawful requirement for a company to actively prevent harm to workers. The second approach, however, is slightly different. Yes, it's focused on preventing future harm from the known hazard, but it's also interested in trying to improve what isn't currently going wrong and thereby potentially preventing anything from going wrong in the future (increasing capacity). Continuous learning and improvement, and not just about what isn't working at that moment in time.

Learning, in the context of New View Safety, involves identifying the "performance adjustments" that your people make daily that enable safe and productive outcomes, the pain points they work with, and the systems or process conflicts that cause friction and frustration. These adaptations and workarounds are invaluable sources of information for process improvement and building resilience, not just addressing actual events. I will cover Learning Teams, a popular approach to organisational learning, later in the book.

EXPERTISE, TRUST, AND INTEGRATION

Where does this lead us then? Well, if our people have to pivot and adjust in order to be productive in our complex socio-technical systems, doesn't this by default make them experts at performing within our organisations? And if we believe this, we must also believe that when bad things happen, maybe, just maybe, it's a sign of deeper issues within the complex systems they use. This assumes, of course, that we don't believe our workers come to work to deliberately have accidents which injure themselves or others! If they are doing this, then you don't have a safety issue; you have an HR and legal issue. (Someone wiser than I said this before, I just can't remember who!) Part of the problem, as I often see it, is that this argument doesn't receive enough consideration, and people involved in incidents are, by fiat, considered reckless, stupid, or worse – uncaring of others!

So, if we are to admit that there will be performance variability, which is highly likely in complex socio-technical environments, like our organisations, and our people are the experts who operate within this environment, then the next logical question is "How do we get our people to tell us all about these variabilities?" Good question. Here's the kicker, though. We can't just demand people tell us when they "break/bend" the rules/procedures. This is especially true if there is a culture of retribution and blame by leadership. As Lloyd (2020) brilliantly points out in his book *Next Generation Safety Leadership: From Compliance to Care*, "trust is the currency of leaders, and bad news only flows where there is psychological safety." If there isn't trust between your leaders and workers, the likelihood that you will gain many insights about performance variability is very, very slim indeed. Why would they volunteer "bad news" if they fear they will lose their job?

If you want to learn more about building trust as a safety leader, I highly recommend Clive's book.

MEASUREMENT THAT MATCHES THE MINDSET

There are many, many people who have far greater knowledge and understanding of safety indicators than I. To name a few, if you want to investigate this topic in greater detail, I would recommend the following (in no particular order):

- Busch (2019) and his book, *If You Can't Measure It....Maybe You Shouldn't: Reflections on Measuring Safety, Indicators, and Goals.* Carsten is known for his critique of purely outcome-based (lagging) safety measurements and calls for meaningful and systemic approaches.
- Dr Drew Rae from Griffith University. Drew challenges traditional safety practices, including the over-reliance on lagging incident statistics.
- Provan and Rae (2026) host a podcast (*Safety of Work*) and have much to add to the cautionary use of lagging indicators.
- Eric Hollnagel, a pioneer in Safety-II (New View) and resilience engineering, has been a leading voice in understanding how organisations succeed under variable conditions and how indicators should incorporate measures of system capacity.

With all that said, allow me to put my view on the following issues with using lagging indicators.

Lagging indicators are inherently a measure of "absence" of safety and don't factor in "presence" that promotes, supports, and enables safety. This attention "void" for leading indicators has been highlighted many times, with devastating events from around the world where the organisations in question held award-winning (lagging) safety performance metrics. A few examples for you to review if you wish, as I don't want this book to be an analysis of incidents from around the world:

BP Texas City Refinery (2005)

BP frequently highlighted its stellar safety performance, as measured by its low number of incidents, before its major disaster, which killed 15 people and injured over 170. The subsequent investigations highlighted an over-reliance on lagging safety metrics.

Deepwater Horizon (BP Macondo Well 2010)

Just hours before this incident, BP and Transocean senior managers were on the oil rig to celebrate the rig's "safety performance" of not having any LTIs for seven years! (I find this highly suspect, as do others!) Injury rates were thankfully low, and investigations describe how the focus on lagging indicators was misplaced. There was also a failure to address leading indicators that are indicative of major incidents such as this.

(Last One I promise!)

Piper Alpha Oil Platform Disaster (Occidental Petroleum 1988)

Occidental Petroleum promoted its low reportable injury and lost time incident rates as evidence of exceptional safety on Piper Alpha. It can be said that this focus shaped their priorities and excluded leading indicators such as adequate maintenance and effective emergency procedures (leading indicators such as the capacity to absorb error). When there was an explosion from a gas leak, 167 workers lost their lives. I was at primary school in the UK at this time, and I remember, even as a 10-year-old, how devastating the event was.

When lagging indicators are linked to performance and literal rewards for employees, the water can get real murky real quick! Evidence shows that when lagging safety metrics are aligned with financial rewards (i.e., actual monetary rewards, which I know can be "serious" amounts of money!), there is a huge incentive to not report incidents and injuries.

Here's something for leaders to consider. I work as a contractor, and like other contractors who work within the field of safety, we are oftentimes told things the workers wouldn't tell the organisation, such as hidden injuries from workplace incidents. This poses a moral dilemma for us as contractors. Do we tell the organisation, or do we respect the trust shown to us by the worker? This isn't a black-and-white situation, as with most situations, it is highly context-dependent. Was the injury sustained due to a systemic failure in high-level controls, with potential for others to get hurt or even killed in the future? Or, was it sustained when a wrench slipped and they broke a finger? These are extreme examples for sure. However, what is important for leaders to consider is the "why" behind hiding them in the first place. What processes or systems are in place which could directly, or indirectly, influence workers to hide such situations?

In one particular instance, the reason was simple from their perspective. If they reported the incident and got sent home on medical grounds, then not only themselves, but the entire site would lose their safety performance bonus as the yearly figures were only days away from confirming their "great safety performance." I won't give the details of their bonuses (needless to say, it was significant!), but I will say this: when they told me how much he would stand to lose, as well as others, I fully understood (and dare I say, agreed with them!) in the actions they took to conceal the injury. Not only was there a (significant) reward coming their way, which they would lose out on, but other workers would also lose out on theirs, and they would know who had ruined their chances of getting their safety performance money! Double motivation to keep quiet!! So, it wasn't just them who were hiding their injury; their workmates (and highly likely others who knew about their injury) were also hiding the injury from management.

Because of the monetary motivation, the organisation was unable to ask the desirable questions like "What can we learn from this incident?" and "Has the hazard/system condition which led to this incident been addressed, or will others fall prey to similar outcomes in the future?"

Does this sound like a culture of safety improvement? What if, instead of paying the safety performance bonuses, the organisation used the money to address all the known hazards effectively and in a co-creative manner with the workers? Sure, the safety stats might not look perfect, but this is what senior leaders need to decide on. Pretty stats, or safer workers! This, for me, is what organisational values are truly about, to guide these types of moral and ethical dilemmas, and not just to be words printed in an annual report or pasted in workers' common areas in fancy font!

(Scan the QR and search my YouTube videos for "Proving safety")

Let's look even deeper. "Blame culture" is a term often thrown around within the safety world. It, along with many other facets of New View Safety, garners polarised views from many who, in my opinion, haven't examined it beneath a surface-level appreciation. So, let's examine "Blame Culture" in relation to my example above.

Blame Culture often refers to the response of the organisation (read "Leaders") after an incident. See above where I discussed Responding vs Reacting, FAE, and O-OP. Given that we already have personal psychological reasoning for attributing blame from the perspective of the way organisations (leaders) choose to deal with incidents, we now also have "self-blame" on the part of the incident recipient (I could prevent myself and others from getting the safety bonus!), as well as blame from others should this person's injury become known to the leaders and safety bonuses are revoked from others also (their stupidity means I lost my bonus!).

Blame is insidious. It's rarely a unidimensional facet that can be addressed with surface-level acknowledgement and management tools. It has its roots and branches across and through many areas of an organisation's culture. Can it, therefore, reasonably be said that lagging indicators (incorrectly used) and safety performance bonuses could be a significant catalyst for a culture of blame? I don't have the empirical data to substantiate this claim, but it would appear so to me.

OK, so we've looked at some of the limits of lagging indicators, and there will likely be many questions, such as "So what do we use instead?" and "Our entire industry is based on these metrics, from winning contracts to reporting on performance! How can we change an entire ecosystem of performative benchmarking?" So, before I proceed, let me reiterate my position before I get a million DM's accusing me of wanting to abolish lagging indicators without reasonable consideration to the implications. I am NOT advocating for the removal of lagging indicators. My position is this:

- Use lagging indicators sparingly and wisely. Who really needs to see them, and what effect might the lagging indicators have, both up and downstream?
- Use them with "other" metrics to give a more complete picture of safety performance. Relying solely on lagging indicators, as we've seen, only provides part of the "safety" picture. Use other metrics alongside to provide context, balance, and "presence" of safety, not just "absence" of safety.

With that said, here are some examples of leading indicators. All, some, or none of these may be applicable to your organisation, as this isn't an exhaustive list. Generally, I suggest to leaders that they get relevant and contextual examples and ideas "from" their people.

Operational Learning and Learning Reviews

Work-as-Imagined vs Work-as-Done (WAI vs WAD) is a popular concept with growing awareness each year, it seems. WAI vs WAD is one of those "perspective" alignments to help leaders, as well as workers, begin to appreciate not just the fallacy of procedural perfection but also the complexity not just of the work, but also of the many factors (resource availability, time pressure, cultural norms, etc.) that can affect and disrupt work. WAI vs WAD can be facilitated effectively via operational learning and learning reviews, such as Learning Teams, another term growing in popularity from the HOP movement.

Learning Teams engage those closest to the sharp pointy end of work to learn about the realities of work, not just what's written in a work instruction. Having been involved in Learning Teams, as well as attending courses provided by here in Australia, I can provide the following caution if you are considering this approach to learn from your frontline workers. Learning Teams in and of themselves are not enough if you want to learn what really happens at the sharp pointy end of your business! There is substantial work, especially with the leaders, that needs to be done first in order for them to be open to hearing the "dirty" realities before you commence with Learning Teams. If the leaders aren't fully prepared for the emotional reactions they will undoubtedly have to deal with when the "truth" is uncovered, there is a very real risk of destroying any further learning from the organisation. Let me explain.

Safety Leader: "We are rolling out Learning Teams, and you are our first group. We are very excited to have you all here. This is a safe space for you to tell us 'exactly' what your world view looks like at the sharp end of operations. Work-arounds, errors, adaptations.... Whatever it is, we want to know so we can improve."

Workers: "Wow, OK, cool, here's the reality.............." (a story about Geoff, a plant operator, and how he had to climb a handrail to reach a shut-off valve because he didn't have time to walk to the supply store to get the appropriate ladder. If the valve wasn't shut off there and then, it could cause a significant issue, and the plant might have to close down)

Safety Leader: Awesome, thank you (reports findings to the senior leadership team, who are appalled by what they are hearing, especially how Geoff violated Golden Safety Rules by climbing the handrail!).

Next day

Senior Leaders: Thanks for being so open and honest with us; that's exactly what we want to get from you all. Keep it up! By the way, Geoff no longer works here as he broke "our" Golden Safety Rules! Keep up the good work, and tell us more!

See what I mean when I say Learning Teams are not enough? You may laugh at the above example, but the unfortunate truth is you only need to spend ten minutes with anyone who has performed Learning Teams in organisations that weren't "ready" to know this is a common occurrence! As Todd Conklin repeatedly states, "You can either blame and punish, or you can learn and improve, but you cannot do both!" If your aim is to continuously learn from your frontline workers (read "experts") about the reality of their work in your organisation, you CANNOT punish them when they tell you the truth.

Systemic Gaps and Post-job Briefs

Systemic Gaps: This is where the safety focus is on identifying performance variability and adaptive capabilities as a precursor to better work, by learning from successful events, as with WAI vs WAD/Learning Teams. Where did we have to adapt? What caused us to veer from our procedures? What can we do/have to account for these, and make work safe in the future?

Post-job Briefs: These are similar to Systemic Gaps, but they focus more on "how work actually happened," and not what should have happened; more of a lightweight version if you like. Questions like the following can be used:

"What went well, and why?", "What was different than expected/any surprises?", "Where did safeguards prove effective or fragile?", and "What should we share or change before the next job?" These post-job briefings enhance learning by focusing on safety as a capacity to be built in real-time.

Regardless of whether you're using one of the above or creating your own, you will see "something" that can be measured quantitatively, as with LTI's. A quick word of caution (again, I know!) when creating alternative measurements; heed the above-mentioned experts' insights, and focus on "quality" of output, not numerical conformance! I've seen companies quantify Learning Teams and say there needs to be "x" number done each month. This new quantitative requirement then becomes the real goal, and as we know, people will find a way of reaching a goal, even if they have to sacrifice quality and substance to hit it! This is classic Goodhart's Law: "when a measure becomes a target, it ceases to be a good measure." So, try to focus on the insightfulness of the activity, not the quantity of it.

AMBIGUITIES AND CRITIQUES

Kool-Aid can be easy to swallow. But is it good for you?

Are we familiar with the saying, whether about politics, artistic productions, or safety methodologies, "They've drunk the Kool-Aid?" For those not familiar, it means the person has bought into an idea so completely that they treat it as the only right way, often without critical scrutiny.

Note: The expression is commonly linked to the 1978 Jonestown tragedy; I use it here only for its modern idiomatic meaning (uncritical buy-in), not to reference that event.

We like ideas that we like; sounds obvious, right? However, just because we like something does that make it an axiomatic truth that isn't open for debate?

A friend of mine who is ex-Special Forces from the UK said of me once, "David, you are very level-headed." If only he knew how I really feel inside!

But this is an interesting point to reflect on here in the book. I've laid out an overview of New View Safety, and if you said my take on it was "biased" towards the use of New View Safety, you would have a very strong argument. So, to even up the scales, and hopefully get ahead of any criticism from people about me only proposing a one-sided opinion about New View Safety, let's look at the opposition's arguments and concerns about it.

Coaching note: A method I use when speaking with coachees one-on-one and they present an uncritical viewpoint about a subject (usually safety), I ask them to complete the following task, commonly called "Consider-the-Opposite":

Write (not type, or dictate, write!) a 1,000-word critique of their current viewpoint. The critique has to be sensible, coherent, charitable (no straw men), and evidence-based. In practice, this means *steelman* the strongest alternative explanation you can, list disconfirming facts/conditions that would make your view wrong, or at least questionable, and outline unintended consequences/risks of your stance.

This can be uncomfortable for people to do. However, evidence shows that once completed, the process can help participants "soften" their axiomatic stance and be more receptive to considering alternative viewpoints.

Now, let's look at some of the critiques and hurdles of New View Safety.

Difficulty Translating Principles into Practical Actions

Safety practitioners can find it difficult to translate the New View concepts and principles into practical actions within their organisations. This is understandable as, ironically enough, the increased complexity that New View looks to address is achieved not through a universal set of tools or defined applications, but by the interpretation of "principles" (both individual as well as organisational) and the very means to instantiate them within an organisation is, well, complex! There are no simple methods or tools for implementing New View Safety; in fact, even the term "implementing" is misleading when discussing New View Safety, as it's not a product that gets "implemented" into an existing organisational structure! This lack of a clear roadmap can be frustrating for individuals and organisations alike.

Perceived Lack of Structure and Control

Traditional safety management often relies on a hierarchical, control-oriented approach with clearly defined procedures and protocols. This structure can provide a sense of security and order. In contrast, New View Safety encourages greater autonomy and flexibility at the frontline, allowing workers to adapt procedures and make decisions based on their understanding of the situation. This shift in control can be unsettling for individuals and organisations accustomed to a more top-down approach and may raise concerns about accountability and maintaining order in complex systems.

And then we have the old reliables!!

Resistance to Change and Fear of the Unknown

A common reaction to New View Safety is one of scepticism, particularly from managers and leaders accustomed to traditional safety and leadership approaches. Changing established practices and venturing into uncharted territory can be daunting for everyone! There is often a fear of unintended consequences, such as creating a "consequence-free culture" or undermining the importance of safety if it is not explicitly tied to punitive consequences for non-conformance! Addressing these concerns and building confidence in the New View Safety approach requires clear communication, strong leadership, and a willingness to experiment and learn from both successes and failures.

Being Poked in the Axioms!

New View Safety is a significant shift in how safety is enacted within an organisation. It is this "significant perceptual shift" that can clash with a person's established beliefs about safety; their axioms get challenged, and that's not a nice feeling for anyone!

> The illiterate of the 21st century will not be those who cannot read and write, but those who cannot learn, unlearn, and relearn.
>
> Alvin Toffler

Life is complex, and to simplify it enough so we can operate in it effectively, we need to establish beliefs and unquestionable truths – axioms. It is these axioms that run automatically and assist us with applying a "perspective" to the world around us. The difficult thing about axioms is that they are mostly unconscious; they lie dormant in our minds quietly influencing our decisions, behaviours and beliefs. If, unknown to you, your axioms are running scripts along the lines of "people should just follow the rules to be safe," "if they deviate from the rules and cause an accident, it's because they don't care about the rules" or even "all accidents are preventable and people need to try harder and be more careful," these can be hurdles to enacting New View Safety principles. Also, imagine you have these unconscious "beliefs" AND you've been rewarded with career advancement for holding them. That's one heck of a self-sustaining set of beliefs on how safety should be enacted, cognitive dissonance in action! There are ways to make these beliefs explicit and begin to challenge them. Welcome to the world of evidence-based coaching!

SO WHAT FOR LEADERS?

So, what does all this mean for leaders in organisations? As already alluded to, if your goal is to embrace New View Safety principles, then the journey begins *within* first, not *with others* first. And by within, I mean within the individual leaders, their perspectives and interpretations about safety in the "organisation." For it is these perceptions that need to be aligned with the principles of New View Safety first, before any journey into New View can be truly effective. As already discussed, with psychological barriers to New View (FAE, O-OP, axiomatic truths etc.), the logical starting point is with individuals who control the narrative of the organisational culture.

This is where we begin our journey. The next section will guide you through the practice and benefits of evidence-based Organisational Coaching as a tool for "individuals," which will then lead into how coaching can support alignment with New View Safety principles. Later in the book, we will then look at how leaders can use coaching techniques in their leadership practice to further embed New View Safety principles and enhance supporting others on the journey to New View Safety. Or indeed, whichever safety methodological approach your organisation is embarking on.

QUESTIONS TO POKE YOUR BRAIN!

- Question 1: When you strip away the corporate slogans, the lagging indicators, and the compliance requirements, what does the word "safety" actually mean to you?
- Question 2: If you accept that success happens far more often than failure, what might you learn if you turned your lens towards understanding how things go right, rather than solely investigating the rare occasions when they go wrong?

Why Coaching, Why Now? 2

WHY COACHING, WHY NOW?

Before I delve into what coaching is and why it's used, I'd like to give some context as to how it came about, and more importantly, how it's developed over the years. And as with all good stories, we have a hero we can reference. The hero of this story is a gentleman called Tim Gallwey; but we will get to him soon.

> I cannot teach anybody anything. I can only make them think.
>
> Widely attributed to Socrates

The foundations of coaching can be traced back nearly 2,500 years to the Greek philosopher Socrates, who developed Socratic questioning as a technique to help people develop critical thinking, uncover assumptions, and engage more deeply with complex problems. At its heart, it sounds simple, asking open-ended probing questions as opposed to preaching or advising, but the effects of this questioning style were (and still are!) profound beyond what you might imagine. People were "encouraged" to examine long-held beliefs, clarify their presuppositions, and be open to considering alternative perspectives. Socrates was of the belief that true knowledge comes from self-examination and reflection, not merely the dissemination and retention of information. Here are some examples of Socratic questions so you can begin to understand how "simple" questions can have a seismic effect:

- What are you assuming here, and why?
- Why do you think that?
- What would be an alternative way of looking at this?

DOI: 10.1201/9781003590538-3

Socrates was admired and loathed. Admired by the youth of Athens and his students, while simultaneously being a pain in the rear to the ruling elites who loathed his "questions" during public debates, especially when they were on the receiving end of them! This just goes to show how powerful questions can be. Socrates was eventually sentenced to death for impiety (essentially questioning too much!), and that was the end of him. But not Socratic questioning, thankfully!

Jump forward a few millennia....

The scientific practitioner model of coaching was developed in 1949 as a tool for clinical psychology, with the Human Potential Movement, which shares its origins with Positive Psychology, coming soon after. And then.....

Remember our hero, Tim? Well, this is where he comes into the story of coaching. Tim was a student at Harvard University in the late 1950s and early 1960s and became the university tennis team captain during this time, as well as being a nationally ranked tennis player. It was during this time that he noticed (presumably while trying to help his team improve) that his team's performance increased naturally and rapidly when he helped shift their focus from self-criticism and rigid adherence to prescribed techniques to self-awareness of their own capabilities and resilience. Call it serendipity if you want, but his approach also aligned with the emerging "human potential movement" of the 1960s, which was a counter-culture movement proposing that humans have vast, untapped potential for growth and improvement. Key players within this field were Abraham Maslow and Carl Rogers, and all three (Tim, Abraham, Carl) recognised the benefits of helping people to realise their "inner abilities."

Tim wrote a book called *The Inner Game of Tennis* in which he states that during a tennis match, every player competes in two games: the outer game (against the opponent) and the inner game (against oneself); and it was this inner game that brought more performance improvements, especially when technical capabilities were already at a high level.

Note: This is an important point I regularly use when I'm trying to explain Organisational Coaching to people who aren't aware of what it is. Usually, people's perception of "coaching" is that of someone who holds valuable technical (usually sporting) knowledge and can share it with someone else with a view to improving their performance (more goals, hoops, etc.). This perception of coaching gets conflated with personal development coaching , and muddies the water! People hear the word "coaching" and assume it's a directive and top-down methodology, i.e., the coach observes behaviours, tells players what to do, and how to do it. However, this approach is limited in its scope, as highlighted by Tim in his book. It's true that most sports coaches do perform this function, but when players reach a high level of competency, there is not much they don't already know about their chosen field of expertise. Consider for a second why the top players in the world (choose whichever sport you like most) still have coaches. Surely if the coach were only imparting technical knowledge, there wouldn't be much they could share with the BEST player in the world, right? For if it were purely a matter of who knows more and has the most experience, then one could reasonably argue that the BEST player in the world would know more! You can apply this analogy to the corporate world also, as Tim eventually did with his work, and say something like "Our people are experts in their field. So how can they improve?" Once people consider this for a second, their next question tends to be "Ah, yeah! So what do personal development coaches actually do then?" Great question. And it's this definition we will look at in more detail throughout this chapter and beyond.

Coaching hasn't rested since our hero moved from the tennis courts to the boardrooms, which he did very successfully. It continues to evolve and adapt to the changing needs of those who rely on it. The "Modern Coaching" era was during the 1980s and 1990s, with the development of some foundational models such as GROW (Goal, Reality, Options, Will), which was first developed by Graham Alexander and then popularised by Sir John Whitmore; and the Co-Active Coaching model developed by Laura Whitworth, which framed coaching as a relationship between the coach and coachee. Based on trust and confidentiality, the relationship is vital for exploring and developing the possibilities and resilience of the coachee.

From the 2000s onwards, coaching underwent another transformation when it was aligned even further with psychological principles and foundational theories. These have legitimised the profession of coaching and placed it directly within the social sciences arena, with the scientific study and application of behaviour, cognition, and emotion. If I were to represent modern coaching evolution succinctly, it would look like this:

- First Generation: Characterised by a problem or goal orientation (the GROW model),
- Second Generation: Characterised by a solutions focus, which included cognitive-behavioural and solution-focused approaches,
- Third Generation: (Particularly associated with the challenges faced within a rapidly changing and hyper-connected world) Characterised by using narrative to focus on meaning-making and reflective capacity.

(There are discussions and propositions about a Fourth Generation, but I think three is enough for this purpose!)

We previously looked at how New View Safety became prominent due to the increasing complexities faced by modern organisations. Well, as you can see, coaching has had a similar journey; evolving to meet the demands of a constantly changing world. Historically, a one-on-one approach, coaching is becoming a go-to investment for organisations all over the world. Here are a few statistics with references that emphasise the growing recognition and appreciation for professional coaching:

- Sixty-five per cent of executives report better decision-making skills and stronger business outcomes after coaching engagements (Entrepreneurs HQ, 2026).
- Seventy-eight per cent of organisations credit coaching with improved employee engagement and retention (Entrepreneurs HQ, 2026).
- The global number of coach practitioners rose 15% since 2023, reaching a record 122,974 in 2025, with industry revenue hitting $5.34 billion USD, nearly double the previous year (International Coaching Federation [ICF], 2025 report).

The obvious question then is, why? What has changed in the business arena that facilitates the adoption and recognition of coaching as a valuable profession? Primarily, businesses are now (and have been for a while) not just locally based with local customers; they need to operate globally. This means organisations now have to develop a global mindset when it comes to competing in a global economy and take into account the complexities that arise from that, such as:

- Adoption of new and advanced technologies.
- Regulatory and compliance complexities.
- Supply chain resilience and flexibility.
- Increased competition and workforce challenges.
- Emphasis on sustainability and social responsibility.

Even if companies are not global players, their level of operational, financial, regulatory, workforce, and social responsibilities is exponentially greater than it has ever been. Think about it for a second – even your local weekend market stall vendor is now connected to, and influenced by, international supply chains, global currency fluctuations, online competition … and on and on. Scale this up to a medium-sized business (20–150 employees), and you can begin to see how complex businesses are. And it is this increasing complexity that is driving the need for a fundamental shift in leadership enhancement, capabilities and approaches, as we will see later in the book. And Organisational Coaching is an empirically validated method for achieving this.

There is now more than ever a need for leaders to tap into their innate creativity, resilience, and self-awareness, and to shift their mindset from command and control to more of an influencing and facilitative approach, due in large part to the increasing scope of stakeholders and employees leaders are responsible for, and the uncertainty that comes along with these. As the old Jewish saying goes, "I wish you many employees," which is known to be a somewhat backhanded blessing. It conveys the idea that having many employees brings many worries and challenges, so it's a subtle way to acknowledge the burden and complexity of managing lots of people and the challenges it brings. Leaders also have limited capacity, spread across many competing facets. It has become increasingly unrealistic for individual leaders to have input/control over all the areas that demand their attention. For this reason, it has become imperative for leaders to enable those around them; to foster collaboration, co-creation, and empowerment. This is the very bedrock of Organisational Coaching, and a coaching leadership style, which we will cover in Part II.

As with any perspective shifts, new skill attainment, or reflection capabilities required to meet these contemporary business demands, they don't just pop into one's head like a computer update. They need to be revealed, discussed, examined, and aligned with updates "by" the individual, usually aligning with organisational requirements. Merely instructing leaders to be more "global" in their thinking, approach, and leadership is not enough, as we saw with changing behaviours in safety.

This is the home of Organisational Coaching.

Note: I'm deliberately using the term "Organisational Coaching" as a way of distinguishing this type of coaching from other forms of evidence-based coaching. Organisational Coaching is primarily concerned with increasing the performance of individuals in the context of an organisation. Other popular terms that refer to this form of "organisational" coaching include Executive Coaching and Business Coaching.

Often, when I'm reading about new or complex ideas, I like to attach my initial understanding to quotes I can easily remember. So here are some quotes from prominent experts on what Coaching is/does. Pick one that resonates with you, write it down, and put it somewhere you can reflect on easily; i.e., a post-it on your work monitor, or, if you're feeling brave, tattoo it on your arm (just kidding, or am I?)

Sir John Whitmore's classic definition of coaching: "unlocking a person's potential to maximise their own performance. It is helping them to learn rather than teaching them – a facilitation approach."

Gallwey (1974), in his book *The Inner Game of Tennis,* describes the benefits of coaching on an individual's performance, which conveniently enough, is a simple quote/equation to remember.

"Performance is Potential minus Interference" is also often written in the following ways (for those of you with a mathematical appreciation): "Performance = Potential – Interference" or even shorter "P = P – I."

Notice the link to "performance" as with Sir John Wittmore's definition? (my personal favourite, probably because it's the shortest one!)

Passmore and Fillery-Travis' (2011) process-focused definition: Coaching is "a Socratic-based dialogue between a facilitator (coach) and a participant (client) where the majority of interventions used by the facilitator are open questions which are aimed at stimulating the self-awareness and personal responsibility of the participant."

When discussing the specialised field of Coaching Psychology (read: evidence-based coaching, which we will address soon), the British Psychological Society [BPS] Division of Coaching Psychology (2022) provides a definition focused on the scientific foundation and application:

> the scientific study and application of behaviour, cognition and emotion to deepen our understanding of individuals' and groups' performance, achievement and wellbeing, and to enhance practice within coaching.

> A coach is someone who tells you what you don't want to hear, who has you see what you don't want to see, so you can be who you have always known you could be.
>
> Tom Landry

> Partnering with clients in a thought-provoking and creative process that inspires them to maximize their personal and professional potential.
>
> International Coaching Federation (ICF)

> Coaching is about facilitating a client's performance, experience, learning, and growth and about actualising goals.
>
> Linder-Pelz and Hall (2008)

If none of those landed with you, try this definition from an LLM; it's a bit messier though.....

> Coaching is a collaborative, solution-focused, results-oriented, systematic process in which the coach facilitates the enhancement of performance, self-directed learning, and personal growth of individuals.
>
> (NotebookLM, 2025)

What about well-known proverbs that help convey what coaching is and the benefits of it? Well, here are two for you to think about:

> Give a man a fish and you feed him for a day. Teach a man to fish and you feed him for a lifetime.

This emphasises the benefits of empowerment and self-sufficiency, as opposed to the provision of a singular output. Coaching 101. I use this proverb when conveying the benefits of both coaching and a coaching style of safety leadership.

(Scan the QR and search my YouTube videos for "Leadership coach reveals professional growth secrets")

> Tell me, and I'll forget. Show me, and I may remember. Involve me, and I'll learn.

This well-known quote (attributed to Benjamin Franklin, but traceable much further back to Chinese origins around 312 BC) demonstrates that active participation in a problem is infinitely more useful for development than the mere provision of an answer. Again, coaching 101.

For those of you who think as I do and benefit from a "bridging" concept from which to start understanding a new concept, allow me to offer this next example. It uses what psychologists call analogical reasoning or mapping and employs a conceptual metaphor as an example. Sounds complex, but bear with me.

Organisational Coaching isn't a great name! It lacks clarity, definition, and purpose, unlike a firefighter, a teacher, or a mechanic. When we think of these professions, we can easily attribute actions, deliverables, and purpose. This isn't true for Organisational Coaching. So, I'm going to build for you the conceptual metaphor I mentioned above, and map it to Organisational Coaching. Here we go!

User-Experience (UX) design suffers from a similar definitional issue as Organisational Coaching. Unless you are a UX designer, you probably think of UX as something along the lines of "making apps look pretty," "making programs easier to use" or "improving usability and engagement." These definitions, though, are just the end product, the deliverables, so to speak. The deliverables of UX design are mostly "felt" rather than "seen"; a better user experience, intuitive process flows, etc. The actual "science" of UX only really becomes clear when it's applied to a contextual situation, such as the following:

- In a start-up business, UX designers may be interviewing customers to gauge how they use the products or digital interfaces provided.
- In a bank, they may be mapping complex workflows to look for streamlined and effective options.
- In a hospital, they may be redesigning patient records to make vital information easier to find in an emergency.

A UX designer will study how users (customers, bank employees, medical professionals, etc.) move through a defined product or process. They then redesign process flows, interfaces and features so it's easier for the users to achieve what matters to them. Similarly, Organisational Coaches study how people move through their work environment. The conversations they have. The decisions they make. The conflicts that impede progress. The goals that matter to them. They then help individuals and teams adjust mindsets, patterns, presuppositions, and beliefs so it's easier for them to achieve what matters to them and the organisation. In this sense, coaching provides a map for the internal and external terrain of individuals, helping them to locate where they are currently (their reality) and plot the most effective route forward towards their desired location (their goal). We will return to this map metaphor in more detail in Part II when we discuss leadership styles that are conducive to enacting New View Safety.

Note: for those of you not already familiar with UX in relation to safety documentation I highly recommend you Google "Usability Mapping."

No doubt some of you will have a priori assumptions about Organisational Coaching, and these may even be causing a degree of cognitive dissonance within you as you made your way to this part of the book; this is very understandable. As I've had many conversations about coaching over the years, I've found it useful to dispel these a priori assumptions as a way of working with a clean slate, so to speak. So let's look at some potential definitional conflicts and define what Organisational Coaching isn't as a way of further understanding what it is.

WHAT COACHING ISN'T!

Below is a table that contrasts coaching with other disciplines as a way of clearly defining what it isn't. It's not exhaustive; it's a quick reference guide to help prevent role confusion and assist with separating it from other well-known roles we see in the world, such as mentoring, consulting (advice giving), and counselling. This section will become more relevant later in the book as we look at Empowered Solutioning and how it logically crosses over some of these divides. But for now, this is a good way of differentiating between known disciplines so as to parse out coaching proper.

RELATED ROLE	PRIMARY FOCUS	HOW IT DIFFERS FROM COACHING
Mentoring	*Experience*	Mentoring often involves imparting experience or advice. Coaching, though, is a facilitation approach focused on helping the person learn, not teaching them.
Consulting	*Solutions*	Consulting involves the professional acting as an authoritative expert who diagnoses and presents the solution. Coaching treats the coachee as the expert in their own life and works collaboratively to construct solutions.
Counselling	*Mental health focus*	Counselling or therapy focuses on coping with, managing, or making sense of the past, often drawing on models to address clinical mental health issues. Pathologies, if you will. Coaching focuses on future behavioural change in a non-clinical relationship.

(Scan the QR and search my website blogs for "Coaching or Mentoring")

BENEFITS THAT MATTER TO LEADERS

As this is a book on safety coaching, I will align the benefits of coaching to the context of safety, and in particular, for this section of the book, towards benefits for individuals. Later in the book, we will look at the benefits of a coaching leadership style, which can be utilised by an individual when leading in a safety context. However, by simply assuming the context of "safety" in the benefits below, one can understand the benefits of coaching in a much broader context, and how it's applicable to every facet of performance improvement, be it work or life.

Goal Attainment

Coaching is frequently used as a goal-focused discipline that sharpens individuals' aims into clear, specific objectives so people can define and deliver organisation-led safety targets more effectively. In practice, individuals set and drive individual (aligned with organisational requirements of course!), actionable safety goals and in doing so maximise safety performance potential. But, before you continue, I'd like to hit home just how important goals are. Goals literally shape our experience of reality. They are more powerful than most people realise. Let me provide a few examples to demonstrate this, as these will be useful later in the book when we look at brain chemistry and individuals' attention and perceptions.

We don't see the entirety of reality through our eyes and then objectively determine what we want to do. This is a popular misconception and a very important fact which is incredibly useful to understand. What actually happens is we perceive reality "through" our internal motivations and perceptions. We already know this; we just don't know that we know it. Here's a popular quote you've likely heard before:

> We don't see things as they are, we see them as we are.
>
> Widely attributed to Anais Nin

In a now classic study, cognitive psychologists Daniel Simons and Christopher Chabris asked people to watch a brief video of two teams passing a basketball and to count the passes made by the team in white shirts. This is a goal-oriented, focus-demanding task given the laboratory setting, desire of the participants to perform well, as well as potential personal motivation to "not fail" (spoiler alert!). Midway through the video, a person in a gorilla suit walks into the centre of the scene, thumps their chest several times, then walks off again. The person in a gorilla suit was clearly visible for several seconds, but despite this, about half the participants didn't report seeing the gorilla at all, thereby demonstrating how intense focus on a goal can cause us to miss even dramatic events directly in front of us. Goals influence what we do and don't see.

This experiment is commonly referred to as "the invisible gorilla" experiment, and their original publication on this phenomenon is Simons and Chabris (1999) published "Gorillas in our midst: Sustained inattentional blindness for dynamic events." If you're annoyed at me for dropping the spoiler for the sake of demonstration here, don't worry, Daniel and Christopher have made several variants to highlight the effect, which are available online, even for people who know to look out for a gorilla. For us in safety, the gorilla has two forms:

1. Not noticing the "weak signal" of potential hazards, as we don't know where to look for it.
2. Not noticing the things that go well because we are too focused on all the things that went badly.

So the science stacks up, but how can this be interpreted into a real-life situation? Well, fairly easily as it happens. Imagine yourself in the following scenario:

You forgot to set your alarm on Sunday night. It's Monday morning, and you have 54 minutes to get into the office for an important meeting with a new client; your work is 50 minutes from your house. You could just make it in time if you rush. Getting into your car, you realise you didn't get fuel on Sunday, as you were relaxing with your family and figured you'd have time to do it Monday morning. Now the morning is here, and you're already tight for time! Your vision and thoughts become focused with only one thing on your mind: getting to the office in time.

You arrive at the petrol station only to be met by a queue of other motorists filling up on Monday morning! Now you're waiting for your turn with everyone else, as you can't risk running out of fuel on the way. It appears like everyone else in the world has ample time to just meander through life! Slowly getting out of their cars to fill up. Wiping the smallest drips of petrol that might have splashed from the petrol tank opening. Reaching inside to look for their wallet or phone while asking the occupants if they would like anything from the petrol station. You are screaming in your head, and maybe a little out loud, "Come on! Get a move on!"

As you watch your car clock tick over yet another minute as you wait, your head fills with thoughts. Thoughts about what excuses you could make for being late. Thoughts about who you could call to buy you more time in the office, and how this would make you look to your new client. Thoughts about what your boss will also think of you for being late. You'll likely berate yourself also for forgetting to set the alarm, as responsible adults should do!

You aren't polite and courteous to the server when you go to pay for your fuel; you're not rude, just need to get it paid quickly! The bridge over a beautiful river you drive over every day no longer provides an opportunity to daydream about going on holiday, as it usually does. Watching the rowers doing their morning training is replaced by scanning for people going too slowly; people leaving gaps in front of them. Now it's just another bit of road you use to get to the office.

The cafe you drive by carefully every morning, as you know, children play while waiting for their parents to come out with their morning caffeine beverage, doesn't even register in your mind. You don't even see the cafe this morning. Cyclists. Elderly people crossing the road. Everyone is now an annoyance to you. They become just another thing that contributes to you being a bad employee; someone who can't arrive at an important meeting on time! You drive faster than normal, yet justify it as being a necessary outcome to avoid blame and self-loathing. You might have stopped at that amber light last week, not today! Today you're on a mission; you have a goal.

Sound familiar??

Setting goals for the future is good. They are positively motivated towards desirable outcomes and states. Having goals for things to avoid, as in the example above, can be just as powerful. Combined, they can be unstoppable.

Imagine having a goal to never feel that anxiety and panic of being late for an important meeting (avoidance goal), as well as a goal to deliver the greatest presentation you can to your new client (positive goal). The world will "literally" reveal itself to you in ways to facilitate those goals. You will notice when your alarm clock isn't set, or even go out of your way to pre-program it and set a backup, just in case. You will prepare for your presentation until you know it back to front, and this confidence will come across in your presentation.

Goals don't physically change reality; they change how you perceive and interact with it now and in the future, which, by default, can change the reality you experience.

Self-awareness

Individuals gain a clearer understanding of how their own biases, beliefs, and emotional responses influence their own and their team's behaviour, and by default, the organisation's "safety" culture (I know this is a contentious term, but you get what I'm trying to say!). Developing self-awareness and self-insight is critical to the self-regulation process, which, as we've already discussed, is paramount when moving to a different approach to safety, as with New View Safety, or just looking to improve the current safety "culture."

Well-Being

Coaching and leadership initiatives can reduce emotional exhaustion and enhance overall well-being by actively managing strain and stress through targeted interventions. Individuals preserve their cognitive capacity for clear, rational decisions in high-pressure or safety-critical situations, thereby lowering risks associated with fatigue and burnout.

Resilience

Coaching can enhance self-resilience among individuals by deliberately activating internal resources, identifying strengths, cultivating skills and psychological reserves, and shifting mindsets. Coaching challenges unhelpful thoughts and uses reframing to adopt adaptive perspectives that open new ways to see, decide, and act on safety.

> When we are no longer able to change a situation, we are challenged to change ourselves.
>
> Viktor E. Frankl

Relationships and Engagement

Coaching helps individuals cultivate listening, appreciation, questioning, and empowerment as they are exposed to it. Witnessing first-hand how these aspects develop during their own coaching experience highlights the powerful benefits brought, not just to themselves, but also how it could relate to others. In my experience, once individuals have gone through an effective round of coaching, they are far more likely to recommend it to others and want to adopt similar skills themselves (and yet ironically, still may feel unable to state exactly what coaching is!).

Unlocking Performance Potential

At both individual and organisational levels, unlocking potential is a consistent outcome of effective coaching. Coaching is a collaborative relationship centred on valued goals and improved performance, with its classic aim to unlock a person's potential so they can maximise their own performance through a process of facilitation and learning, rather than teaching.

You might finish reading the above benefits and reasonably think something like "Well, they all sound awesome David. But how can you confidently state these benefits are replicable?" or maybe even "Great, but it sounds a bit too good to be true! Is there any evidence/science to back up your claims?"

These are good questions, and something we will look at next.

WHAT "EVIDENCE-BASED" COACHING MEANS

Say, for example, you need a mechanic to fix your car, or an accountant to do your tax returns. You would definitely (and rightly so!) require any individuals you were considering to have their skills aligned with proven industry standards, in whichever guise this comes; professional development via a certified training provider, tertiary, higher education, etc. You wouldn't, for example, be willing to let a mechanic whose training only came from watching their uncle tinker on his hobby car work on your car, especially if your car is expensive! The same can be said for coaching. Lots of time and money are put into employees at companies, and hiring a coach who isn't trained in rigorous coaching principles and practices could be highly detrimental to the company and the individuals!

However, here's the rub – coaching per se is largely unregulated. Even a casual glance through a Google search of "coach training" will produce a plethora of available courses, lessons, certificates, and other offerings. Unlike with the mechanic or accountant, anyone can put the word "coach" on their resume or on their business cards with no reprisal, professional accreditation dispute, or legal consequences. This poses a HUGE problem for individuals and organisations alike. How do we know the coach we are using is actually competent, credible, and capable of doing what they say? How do we know that the techniques they are utilising are actually proven and aligned with best practices?

Institutions such as the International Coaching Federation (ICF), the European Mentoring and Coaching Council (EMCC) and other such bodies have made huge strides towards professionalising coaching, with strong industry alignments, benchmarking of skills/qualifications, etc. However, the issue still remains; the word "coach" is still unregulated, and as we've already discussed, equally hard to differentiate the word from all of its actual and perceived uses. This brings us nicely back to the heading of this section – "What evidence-based coaching means."

(Scan the QR and search my YouTube videos for "Coaching Integrity")

To help with this section, I've created a visualisation to help explain evidence-based coaching (EBC) and the important role it plays:

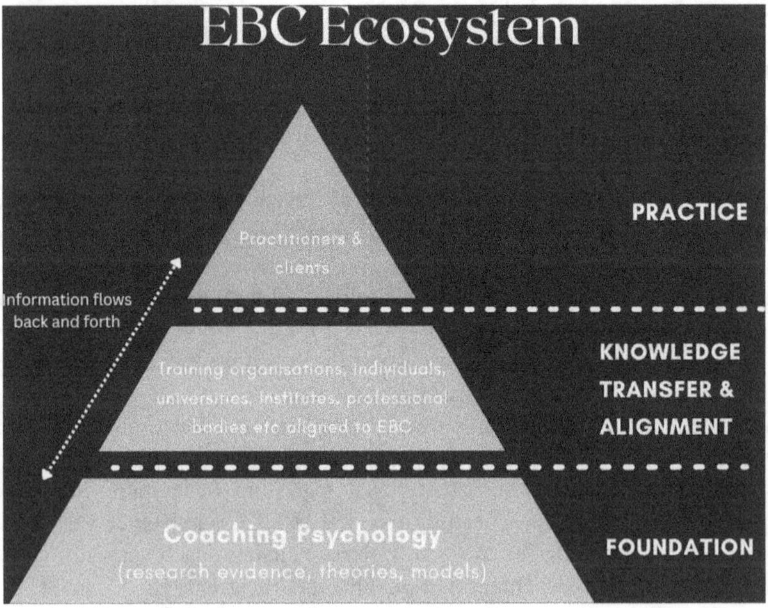

We can see that the job of evidence-based coaching (Coaching Psychology) is to form the foundation upon which all "upstream" knowledge transfers and practices should be based. Why would it be any other way? If we want a profession to be legitimate (evidence-based), then it's required to be built on empirical, peer-reviewed, science-based theories, professional expertise, and rigorous evidence. Only then can we base practices and techniques on validated methodological frameworks, backed (as best we can at any given time) by scientific scrutiny (repetition, validation, quantifiable evidence, etc.).

Let me say it this way.

If I were a leader in a company, or an individual looking to hire a coach, I would want them to be at least trained by a respected and professional registered training organisation (RTO) which aligns with known methodologies (GROW, PEAK, etc.) and ideally aligned with a professional standards body. Similarly, my requirements for a safety coach would be qualifications and experience in safety, and an awareness of how coaching skills can be utilised in a safety context. This could come from specialist training providers or independent qualifications in both disciplines. Remember, with professional RTOs or specialist training providers, you get what you pay for, so expect to pay decent money for a decent training course and certification programme; not the "$12 to be a fully qualified coach in 30 minutes," of which there are many advertisements for! These only degrade and devalue the profession.

As with all purchases in life, "caveat emptor" (let the buyer beware!).

COACHING ACROSS LEADERSHIP LEVELS

Context is king. Not just to safety, as we've already discussed, but to almost every complex situation we will face in life. The vagaries and variables present all around us mean that we **must** factor in context.

And the same can be said for coaching approaches.

Only having and using one coaching approach or framework for all individuals within an organisation doesn't factor in the reality of the individual within the context of the organisation. The hammer and nail metaphor comes to mind here. If all we have is a hammer, then all we see are nails! And it is for this very reason that there are multiple modalities of coaching which are more contextually relevant to different levels within an organisation. If we think about it for even a second, this makes perfect sense! A frontline leader is more concerned with, and therefore motivated by, the performance of certain individuals within their limited sphere of control and influence on a day-to-day basis. Whereas an executive within the same firm will have a more holistic set of goals and the associated motivations, due in large part to their extended sphere of control and influence. They are less concerned with the micro-decisions of individuals and more concerned with how people en masse lead, innovate, and affect culture.

An important note to make here. Although the proposition I laid out above is generally accurate, the modalities available to coaches are not mutually exclusive to certain roles within the organisation! What do I mean by that? Let's consider the following. Senior leaders, for example, can be more concerned with and motivated by the organisation from a low-resolution (holistic, high-level) perspective; this doesn't mean that they don't have challenges with direct reports, or even their own personal performance within the organisation. At the opposite end of the leadership spectrum, we have a similar situation. A frontline leader who is generally more concerned with the micro-decisions and behaviours of their direct reports might still benefit from a more holistic modality of coaching as a way of motivating their direct reports to the "why" of the organisation, as well as their own personal performance enhancement. To add even more complexity, some coaching modalities are specifically designed to cater for multi-level benefits, individual and organisational. But even these "encompassing" modalities can be complemented with the use of other modalities!

We can quickly see, therefore, how important it is for coaches to have multiple "tools" at their disposal in order to effectively meet the contextual requirements of the individual. This is known as "meeting clients where they are." And it is for this reason that this book is separated into two distinct parts: (Part I) coaching for individual performance, and (Part II) coaching-based leadership; coaching techniques employed by the individual as a safety leadership style.

Let me try to make this more concrete by highlighting the distinctions of some of the approaches available to coaches. The table below is a condensed overview of the complexities that I've already covered, and even a cursory look through the evidence-based coaching literature will highlight many more nuances, adaptabilities and combinations available to trained coaches.

I've segmented it into just two (extreme) roles within an organisation for ease of comparison: Executive and Supervisory. And to keep with the theme of this "part" of the book, I've kept it to the "individual" perspective, as we will touch on the "leadership" aspects later in the book.

This table focuses on an executive's individual development.

COACHING MODALITY	RATIONALE	COMPLEMENTARY MODALITIES
Cognitive behavioural coaching (CBC)	Excellent for addressing performance interfering thoughts (PITs) such as perfectionism, procrastination, and anxieties that block individual goal attainment.	Motivational Interviewing (MI), which is useful if the executive is ambivalent about committing to personal change or development.
LLG (lead, learn, grow) model [a]Generally seen as an "almost" all-encompassing coaching modality for leaders	An aspect of LLG addresses personal practices, continual learning, and goal attainment.	The SPACE (social, context, physiology, action, cognition, support) Model, which can be especially useful for high-stress situations such as public speaking.

Note: [a]There are many more variables and combinations available for differing requirements.

This table focuses on a supervisor's individual development.

COACHING MODALITY	RATIONALE	COMPLEMENTARY MODALITIES
GROW & RE-GROW models	Foundational in providing structure and momentum to individual performance goals.	PEAK model is excellent for specific performance factors related to ability and know-how (Task performance = Ability × Effort/Difficulty).
Rational coaching (RC) model	Addresses performance interferences directly by challenging rigid demands (low-frustration, tolerance, etc.) and is especially useful in roles where output is closely scrutinised.	Techniques of CBC, such as the "Responsibility Pie," which help the supervisor reattribute responsibility when they tend to blame themselves or others, disproportionately for outputs.

Note: [a]Again, there are many more variables and combinations available for differing requirements.

As you read through the above tables, you may have found yourself thinking, "I can see how these examples could cross from one role to another!" And you'd be right! They can, and do. The tables, however, provide a low-resolution, heuristic (rule of thumb) approach, and as with all situations, it will depend on the context of the individual. I told you coaching can get complex quickly!

Here is a non-exhaustive list of some available evidence-based coaching modalities:

1. GROW Model (Goal, Reality, Options, Will).
2. RE-GROW Model (Review, Evaluate, Goal, Reality, Options, Will). With many other variants, including I-GROW, T-GROW, etc.
3. ACHIEVE Model (Assess, Creative, Hone, Initiate, Evaluate, Valid action, Encourage).
4. OSKAR Model (Outcome, Scaling, Know-how, Affirm/Action, Review).
5. CIGAR Model (Current situation, Ideal outcome, Gap analysis, Action plan, Review).
6. LLG Model (Lead, Learn, Grow).
7. Rational Model (focuses on disputing four types of irrational belief, such as demands, awfulising, frustration tolerance, and depreciation).
8. SPACE Model (Social context, Physiology, Actions, Cognitions, Emotion).
9. Solution-Focused Coaching (focuses on constructing preferred future states, noticing exceptions, and leveraging existing resources).
10. Motivational Interviewing (used to resolve ambivalence to change, elicit intrinsic motivations and often draws on the Transtheoretical Model of Change).

Modalities can be segmented by their "strengths" or "capabilities," as highlighted in the table with the Executive and Supervisor examples. Grouping of issues and specialist focus capabilities are shown in the table below:

ISSUE ADDRESSED/FOCUS	MODEL/STYLE NAME
General goal and iterative goal attainment	GROW & RE-GROW models
Comprehensive problem-solving	PRACTICE model
Systemic change	LLG model
Goal planning	ACHIEVE model, logical progression from the GROW Model
GAP analysis/strategy	CIGAR model
Psychological blocks/cognitive rigidity	Rational coaching

ISSUE ADDRESSED/FOCUS	MODEL/STYLE NAME
Emotional regulation	SPACE model
Performance enhancement	PEAK model
Solution construction/strengths focus	ENABLE model (solutions-focused coaching principles)
Team wellbeing	CBTC (cognitive behavioural team coaching)
Habit change	BIG (behaviour incompatible with goals) Model
Change management	Transtheoretical model of change

Regardless of the Model or Style used, there are overarching ethical requirements. Think of them as "rules of engagement" or "laws" of coaching. It is not morally acceptable to utilise psychological tools without these ethical guidelines, just as it's not morally (or legally!) acceptable to drive a car without the prerequisite training, adherence to laws, etc. Both outcomes can have disastrous implications if the "rules" aren't followed to the best of one's abilities.

In the next section, we will look at the ethical requirements and obligations of professional coaches.

ETHICS, SCOPE, AND WHEN TO REFER

Professionals should abide by professional standards, ethics, and guidelines. Why? The mere accumulation of information through training is not enough; there needs to be guardrails in place to regulate behaviour within the profession. We have them with almost all other professional roles in our society: law (National Bar Associations), healthcare (International Code of Medical Ethics), and finance (International Code of Ethics for Professional Accountants), to name just a few. And if professional coaching is to be considered a "professional" role (as it should be!), it too needs to have and reference a set of professional standards and ethics. This is such an important issue that it was given an entire semester during my studies at the University of Sydney.

For professional coaching, there are several entities that have made the biggest strides towards professionalising through codes of practice, ethics, values, and expectations for the professional practice of coaching. The International Coaching Federation (ICF) is currently at the top of that list, with notable and influential impact from the Association for Coaching (AC) and the

European Mentoring and Coaching Council (EMCC), which jointly created a global code of ethics.

Regardless of the origin of the ethics and standards you prescribe to, there are several commonalities that they all share and are considered "essentials" for professional coaching. Here are a few:

Contracting

Think of this as a contract (obvious, I know!) between the professional coach and the client (coachee). It will cover such items as financial arrangements, roles and responsibilities, confidentiality (this can become difficult to negotiate if an organisation is paying for employees to have professional coaching!), and even methods of professional coaching used. As we have seen, there are many methods for professional coaches.

Referring

The key to this aspect is recognition of the definition of professional coaching. During a discussion with Charity Becker on my YouTube channel, we spoke about this, and I think the easiest way to convey its intent is as this. Professional coaches are trained to work with individuals and teams for the improvement of potential, essentially (as we discussed earlier with the definitions of coaching). They are <u>not</u> (generally) trained mental health practitioners, such as mental health counsellors, clinical psychologists, or psychiatrists, who are trained to work with people who have diagnosed or suspected mental health conditions or disorders such as generalised anxiety disorders, depression, or PTSD; and unless the professional coach also has such qualifications, they should not exceed their scope of practice and limits of competence, and refer the coachee appropriately.

That aside, it also refers to interpersonal compatibility. Sometimes coaches and coachees just don't "connect," for whatever reason, and the coach must be humble and concerned with the coachee's development enough to be able to recognise this and look to refer them to another professional coach who may be a better "fit" for the coachee.

Psychological Safety

This covers a wide array of facets from confidentiality (again) and cultural sensitivity, through to secure data handling and appropriate interactions between the coach and coachee. The core tenet, though, is one of trust; as without trust, there isn't a coaching relationship! At best, you might have a transactional interaction (an advice-giving relationship); at worst, it can lead to learned helplessness by the coachee (over-reliance on advice from the coach, and thereby disempowering them), or even ethical breaches and abuse of power by the coach, i.e., using confidential information to manipulate! The importance of trust cannot be overestimated here. Are you likely to discuss your inner fears or goals with someone if you don't trust them? Probably not. And if you're not discussing important issues, what is the point of coaching?

As with professional coaching, these tenets, codes of ethics, and practice are also relevant to safety coaching and should be inherent in the practice. The contracting may be implicit within a larger (commercial) contract, and the advice giving may need contextualising (we'll discuss this soon with Empowered Solutioning); however, the intent and practice of safety coaching should still abide by professional standards and code of ethics as much as possible. Speaking of which, let's take a closer look at coaching and how it perfectly aligns with New View Safety approaches.

QUESTIONS TO POKE YOUR BRAIN!

- Question 1: Considering Tim Gallwey's equation that Performance equals Potential minus Interference, what internal "interferences" might be currently clouding your ability to unlock your full potential?
- Question 2: In your daily interactions, where might the desire to be the "expert" who provides the answers be overshadowing the opportunity to be the "facilitator" who helps others fish for themselves?

Why Pair Coaching with Safety? 3

WHY PAIR COACHING WITH SAFETY?

There is a gap. A gap between understanding New View Safety and translating these principles and theories into actionable, day-to-day behaviours. I know I'm not the only one who thinks like this. Even a cursory look through LinkedIn will support my hypothesis. And as we've already discussed, New View Safety isn't a heavily defined or prescriptive safety methodology; it's a set of values, assumptions, and "new" ways of perceiving safety. If we know this is true, which it is, why are we surprised when people agree in principle with New View Safety, yet are confused when it comes to what it "looks" like in their organisations? As our prior hero Tim might say, the lack of clear operational action is the "interference" to the "potential" (knowledge) and the "performance" (New View Safety enacted).

Merely knowing something is right or good isn't enough!

Are we missing the bridge between the theory and practice of New View principles? I don't believe so. I believe safety coaching is the missing bridge between knowledge and action, between theory and practice. It connects New View Safety principles and values with repeatable, real-world performance, built on clear psychological foundations that make them livable, and not just aspirational or theoretical.

As a professional practice, coaching has more than enough evidence to support its efficacy. Its foundations and alignment to evidence-based psychological practices and principles ensure this. As you will have seen in Chapter 2, there are dedicated coaching approaches for a myriad of situations, from change management, goal attainment, to performance improvement and emotional regulation, to mention just a few.

Let's look at how the practice of evidence-based coaching can assist with enabling movement towards a New View Safety approach.

DOI: 10.1201/9781003590538-4

NEW VIEW SAFETY VIA COACHING

By unlocking people's potential and providing a structure for relationships and communication, evidence-based coaching supplies the conversations and mindsets that make New View Safety workable. Below are some "interferences" that can surface when moving towards New View Safety. Coaching can directly address these, thereby facilitating the move to "workable."

(Scan the QR and search my website blogs for "Synergistic beyond belief")

INCREASED SELF-AWARENESS

Increasing self-awareness is coaching 101. Why is self-awareness so important, though, especially when we are discussing safety? The short answer is this: an increased self-awareness helps individuals recognise internal interferences (remember the equation from earlier in the book *Performance Is Potential Minus Interference*). It therefore becomes highly beneficial for individuals, whether they are moving towards New View Safety principles or not, to learn more about themselves on a deeper level, especially during safety-critical situations.

> You have power over your mind, not outside events. Realise this and you will find strength.
>
> Marcus Aurelius

A heightened self-awareness can help with recognising internal interferences that provoke a "reaction" (and let's assume for this example that the reaction is negative in relation to the desired "safety" outcome) and facilitate a more considered and safety-aligned "response." This shift from reaction to response is a cornerstone of many New View modalities, with HOP having it as one of its five core principles: "Response Matters." And remember this about HOP: it's not merely a safety methodology; its core purpose is Human and Organisational Performance, with improved safety outcomes being one of many benefits, or side-effects if you will. So we can see that by raising self-awareness in individuals, there are many benefits associated with it, not just safety!

Sounds good, right? Helping individuals recognise how their psychological "interferences" might be influencing how they react to stressful situations. But how is it done, and what is the individual to do with this information? This is a good question and one we'll look at now.

Awareness leads to examination, which leads to evaluation and beneficial adjustments.

As with most psychological interventions, the first step is the most important: the recognition of a current psychological state by the individual. Without this, there can be no improvement. Once the recognition has taken place, there then needs to be a cognitive process of evaluation, followed by a more rational and informed decision of how to act (respond). Put simply, this is a process a coach could go through with an individual:

1. Trigger (noticing one's own emotional state)

Recognition that they are experiencing a shift in attitude, emotions, desires, etc.

2. Pause (engaging reflection and insight)

This "pause" allows deeper self-reflection and analysis. What was it that triggered me? Am I operating on internal biases? What am I assuming here? Essentially, what happens is an internal objective analysis, which allows the individual to examine the thinking and feeling that triggered the desire for the immediate "reaction."

3. Choose (which response is most appropriate given the desired outcome)

Once the individual has recognised the trigger, objectively analysed any internal "interferences" (axioms, biases, frames of perspective, etc.) that were the catalyst to that reaction, they are then able to make more values-based and considered decisions on how they "respond."

This process isn't a one-off exercise. Far from it. It requires the coach to hold the individual accountable and facilitate and support continuous self-reflection and learning of the individual. For this type of coaching, the RE-GROW model is particularly useful as it includes the elements of "Review and Evaluate" into the already well-established GROW model; thereby allowing continual improvement over many sessions towards a desired future state, while taking into account proximal contextual factors and allowing for adjustments along the way.

We've covered a lot so far in this part of the book: what New View Safety is and isn't, what coaching proper is and isn't; and the symbiotic relationship between the two. For the last section of this part of the book, we are going to be looking at how they come together in real life. We will be looking at research, my experience as a safety coach, and explaining some of the nuances of the practice of safety coaching. This is where the rubber hits the road.

OVERCOMING INDIVIDUAL INTERFERENCES AND BIASES

Earlier in the book, we discussed several biases that are, I believe, commonplace when people are asked to operate within a New View Safety paradigm. These were:

- Fundamental Attribution Error.
- Confirmation Bias.
- Hindsight Bias.
- Other-Orientated Perfectionism.

Fundamental Attribution Error and Other-Orientated Perfectionism

Safety biases manifest most when things go wrong. When mistakes are made. When incidents happen. The result is that we blame people's behaviours or actions on their actual (or perceived) internal characteristics; he's stupid, she's accident-prone, how stupid can one person be, I expect better from them, etc., instead of looking at the broader systemic influences involved, such as constrained resources, time pressure, task conflicts, system misalignments, etc.

Coaches will first and foremost actively listen (which is not as easy as it sounds!) and ask questions for the coachee to explore their reasoning, as well as "potential" other factors that could have contributed. Questions such as "What's another way of explaining the person's actions?", "How might the other person describe the situation?", and (this one has a sting in it!) "What was your role in creating the situation?" With questions such as these, at the right time, and repeated examination and self-reflection, the coachee can begin to raise awareness of their automatic assumptions.

Hindsight Bias

This bias is well documented, as with FAE, and isn't confined to the world of safety, as with all biases we will discuss. You'll likely hear statements from others (or yourself!) such as the following, which are a good sign Hindsight Bias has reared its ugly head: "Why didn't they just go that way?", "How did they miss that sign?", "They shouldn't have done X, they should have done Y instead. It's obvious!" Hindsight Bias is a comfortable stance to adopt. We get to feel morally superior and intelligent, all in one fell swoop! The problem, though, is that if we default to Hindsight Bias, we don't actually get to learn anything about what "actually" happened and why! All we have is a feeling of superiority, "knowing" we would have done it better/differently, and the person involved should have known better!

Again, as with the Fundamental Attribution Error, the objective is to provide a space for coachees to walk through their thoughts, assumptions, and axioms, and begin to become aware of them. This is a common theme throughout coaching. People generally assume that because something came to them, a thought or feeling, it must be true and accurate. It's only when they begin to examine these presuppositions that they begin to realise how they have more choice in how they act and react.

Coaches will likely challenge the coachee's perspective, thereby encouraging critical thinking, and germinate the idea that there "may" just be a better way to view incidents. A way where we don't act on our biases, but we recognise them when they surface, acknowledge them, but don't allow them to influence our actions. There's no getting away completely from biases, especially if we've had them for a long time. Awareness, though, goes a long way to controlling them. I knew a safety leader once (a former coachee of mine) who would appoint a member of his investigation team to be the "bias police." Their role was to listen for biases from other members of the investigation team and call them out. A fabulous idea he had in a coaching session, once he realised the power biases have on incident investigations.

(Scan the QR and search my website blogs for "Biases in safety leadership")

Cognitive Dissonance

Cognitive dissonance is when we hold two or more competing concepts, beliefs, and/or values. The result is psychological discomfort, which can hinder or even completely prevent someone's ability to fully embrace new concepts and values (as we have already discussed are part and parcel of New View Safety), especially in the face of operational pressure. This is a crucial issue to address, as New View Safety often introduces philosophical concepts that can conflict with people's existing, and often deeply ingrained, beliefs about safety. And let's be honest, it's easy to see why, right?

So, the question again is, what can coaching do to help with this "interference?"

First and foremost, and this may seem obvious, coaching helps people articulate that they actually have conflicting beliefs. People generally know what's required of them, yet frustration can set in when it doesn't come to them when they need it the most. Through Socratic questioning, coaches can help coachees understand what the "interference" is, which is a relief in and of itself. As we looked at earlier, questions such as "why do you believe that?", and "what assumptions are you making here?" can help the coachee explore their thought patterns and begin to challenge their own assumptions (axioms, presuppositions, etc.) and begin the process of considering alternative perspectives.

> Progress is impossible without change; and those who cannot change their minds cannot change anything.
>
> George Bernard Shaw

Aspects of Cognitive Behavioural Therapy can also be employed in what's called Cognitive Behavioural Coaching. This approach helps the coachee identify and modify their own counterproductive thought patterns and replace them with more balanced and constructive perspectives that are aligned with New View Safety.

REFRAMING OUTCOMES FROM "NO INCIDENTS" TO "CAPACITY TO ABSORB ERROR"

This concept of "capacity to absorb error" is a central theme to all New View methodologies and is linked directly to principles such as "error is normal" in Human and Organisational Performance (HOP). It's one of those fundamental principles that can be difficult to grapple with. The shift from aiming for no errors to be made (Old View) to acknowledging that errors will always be made and therefore the systems must anticipate and cater for these errors (New View) can be done through reframing what good, or successful, looks like. By examining structured conversations, what emotions and thoughts are present for the coachee, the coach can help them reframe their perceptions and goals from a purely "preventative" frame, with its potentially associated negative emotional state, to a state where there is value in learning from previously negative consequences, as well as anticipating future errors and making plans to mitigate them.

I had a coachee once who got angry every time he heard of an incident within his team. He explained how the anger consumed him and affected how he managed the incident, mostly in a negative way. After a few coaching sessions, he came to the realisation that his anger came from fear. His fear came from a good place; he cared about his people, but all his team saw was his anger. During his coaching journey, he learned where his anger came from and decided to reframe his perception to one of "the price is already paid, what do I have to learn to prevent others from experiencing the same in the future." This reframing helped him manage his anger. It prepared him for adopting a curious mindset and learning. It reduced his level of anger considerably. As a consequence of his own reframing, his teams began to see his innate care and compassion, as well as his professional desire for safety improvement, as their leader.

**The above example is condensed and paraphrased for ease of highlighting the capabilities of reframing.*

DEVELOPING CURIOSITY AND HUMILITY

Curiosity is the antidote to blame and the doorway to learning. Humility is being willing to learn. These are vitally important skills to learn if New View Safety is to even stand a chance.

You may have noticed, as we discussed some of these biases, that the coaching approach used to help coachees address them is a form of curiosity and humility by the coach. Socratic and probing questions, and as with coaching proper, a deep humility on behalf of the coach so as not to assume they have all the answers, thereby allowing the coachee to develop their own understanding and associated solutions. In my experience, once leaders in particular experience this first-hand in a coaching session, they themselves become curious about coaching skills. They've exposed more of their own thoughts and unexamined biases, created solutions by themselves where previously they may not have been able to, and experienced the empowerment that comes from the process; and they see the benefits of this for their own leadership style.

We will address this further in Part II, where we look at coaching as a leadership style.

THE REAL TENSION

Ever since I was first introduced to coaching in a safety context (through a forward-thinking safety consultancy firm here in Perth, Western Australia called "Barclayss"), I've had discussions about its application. The debates with my directors generally went along these lines:

Me: Coaching properly doesn't allow us to give advice or instructions.
Director: Sure, but sometimes we have to!
Me: So how do we reconcile these two competing needs?
Director: We decide when to use which approach because we are there to improve safety for the client

This is an abridged version of conversations that would take place. You can see, though, that the issue is one of competing priorities: service delivery, client expectations, and professional practice. To add more confusion to the mix, it was also a definitional problem. What exactly is a safety coach, and what can clients expect them to do?

Conversations such as the one above have been in my mind ever since, probably due to the confusion I experienced while I began my journey into safety coaching several years ago.

How can we give advice (consult) and coach, professionally and ethically? How and when do we switch between the two disciplines?

This conflict and confusion was the birthplace of the idea of Empowered Solutioning for me. I originally conceptualised it as a combination of "consulting" and "coaching." However, over time, and with more reflection, I have broadened it to a more general concept of "advice giving and coaching within a safety context."

From my personal experience, I can attest to the variability in the desired capability of safety coaches. The most relevant situation I was involved with was a project on a mine site here in Western Australia. The client wanted leadership-based safety coaching, and this was to include "walk-about" with the leaders in their specific area of responsibility on the mine site. Sounds simple enough, but when we look at the distinct approaches listed above, you can hopefully see the dilemma with this. I would start with the leader in their office (personal development), go into the field and walk around with them and be expected to switch between system diagnostics (tools, processes, and environment) and observation on behaviours, depending on what we were witnessing. All the while trying to engage the leader with questions to facilitate personal development. The cognitive load needed to keep track of what I wanted to achieve, how, and which approach to use was draining, to say the least! This most definitely reduced my capability to be as effective in any one of these approaches. In hindsight, a more distinct and separated approach to the leadership safety coaching would have been better for me, and likely for the leader I was engaging with!

For those of you who regularly coach, you will hopefully be able to relate to the situation I was in: switching between purely questioning, guiding for a desired outcome, advising, and back and forth multiple times! I didn't really enjoy this project as my role wasn't clearly defined by context. I was performing mental gymnastics, and I'm no gymnast!

So what's the solution to this?

At the time, I was struggling with the dilemma of questioning, systems diagnostics, and observing behaviours; I was unaware of the EMCC (European Mentoring and Coaching Council). If I had discovered them sooner, I might have been able to write this book sooner! Or maybe it would never have been written at all, as they are addressing a similar issue – albeit not in a safety context. Here is a summary of their purpose from their website:

> The EMCC exists to develop, promote, and set the expectation of best practice in mentoring, coaching, and supervision globally for the benefit of society.

Their framework expects practitioners to draw on a range of approaches, which in practice includes knowing when to be more facilitative (questioning) and when it may be appropriate to share expertise and knowledge (systems-diagnostics/observation). See the similarity with my dilemma? There are obviously guardrails for transitioning between these roles, similar to those found in coach training, and these include things like:

- Clear definition and contracting of primary function (again, this may be difficult for safety coaching, especially if you are in a situation like the one I highlighted!).
- Seeking permission before switching between roles (this is fundamental for building a relationship based on trust).

Jonathan Passmore is closely involved with EMCC as an accredited practitioner, adviser and research collaborator. This is important to know, as Jonathan is also one of the major contributors to peer-reviewed articles on safety coaching, and his work, as well as others', is a major influence in the research of safety coaching.

THE THREE APPROACHES TO SAFETY COACHING

Ok, so now I've given you some context as to the motivator for the idea of Empowered Solutioning (the subtitle of this book), as well as other movements around the world that are grappling with similar issues of "coaching" and "advice giving," let me begin by laying out what Jonathan and others have discovered in their research into safety coaching.

First and foremost, the evidence for safety coaching being effective in high-hazard industries is undeniable. What isn't so clear is the definition of the term "safety coach." Here are three options highlighted in the literature on safety coaching. I have inserted terms used in my story to help connect my situation with what the research highlights.

Safety coaching is, or should be, one of the following:

1. A *Socratic-focused* dialogue which focuses on the individual internal motivation, biases, goals, and "interferences" of the coachee. Results can be both proximal (here and now) and distal (future focused), depending on the coaching requirements of the coachee.
2. An *analytically focused* process which focuses more on high-hazard potential events, the environment, and the systems the employee is using and working within. It focuses on mid- to long-term future operational states.
3. A *behavioural view* which has a proximal focus on actions of employees, and sometimes associated with Behaviour-Based Safety (BBS), which, for those of us on LinkedIn, is one of those topics that generally garners strong opinions!

Here's a simplified table to show how this looks in practice:

APPROACH	SAFETY INSTRUCTION/ ADVICE	COACHING INFLUENCE	CURRENT OR FUTURE FOCUSED
Socratic dialogue	Minimal, if any (although an awareness of safety, especially from an organisational perspective, may be useful)	High	Can be both, but predominately future focused.
Analytical	Moderate	Moderate	Mid to long-term transformational safety targets. Data-patterns-context.
Behavioural	High	Low	Here and now, observable.

The literature discusses pros and cons of each approach, and to highlight these, here is yet another simplified table for you:

APPROACH	PRO'S	CON'S
Socratic dialogue	Highly effective for individuals to reduce "interference" associated with a desired safety approach at a deep level. Facilitates safety beyond surface-level compliance.	Hard to measure and replicate to exact standards. Subtle and gradual. Takes time. Dependent on the coach's personality traits. Role confusion as to what the safety coach is actually there to do.
Analytical	A key component of leader-based intervention programmes. Recognises the complexity of systems. Quicker with more measurable results compared to the Socratic method.	The intervention must be embedded across all aspects of the organisation to facilitate change. Limited impact on companies already performing well. Doesn't directly address how safety needs to be communicated in order for it to be sustained.
Behavioural	Has been associated with significantly reduced incident and injury rates. Immediate interventions that are easy to measure. Behavioural interventions can have lasting effects for months.	Often linked to BBS. Focus more on prevention than resilience and capacity. Potential to be seen as "box ticking" and even as a punitive exercise, if not managed properly.

If neither of those tables resonated with you, let me try another way, with another table (of course!), with some example "engagement" statements/questions to help understand how it differs.

APPROACH	ACTIONS	ENGAGEMENT
Behavioural	Observe critical steps, engage, and give feedback.	Show me that step. What's your critical control here? One thing I noticed. You missed this step. Do this instead.
Analytic	Apply the approach with consent.	Where's the weak signal here? How could you build in more resilience? Let's map out a bow-tie with triggers and actions.
Socratic	Engage in a coaching contract. Goals, reality, options, and will (GROW, as an example framework).	Where are you now? Where would you like to be? What do you have, or what could you create, to improve safety by 10%? How important are these to you?

Notice in the table above that as one moves from "Behavioural" to "Socratic," the questioning increases.

So there we have it. The current debate has isolated safety coaching down to these key approaches and functions. But, if, like me, you looked at those and thought something like, "I can see use and benefits in all of those approaches, it just depends on the context of the organisation, stage of the project, or role the safety coach is playing," then we are on the same page (literally, as well as figuratively!).

Looking at the three defined approaches, we can clearly see there is always a "need" for the safety coach to have safety knowledge, to some degree, whether it's technical and specific (as with the behavioural and analytical approaches), or whether it's more holistic and organisation-based (as with the Socratic approach). I must caveat that last sentence with the following. The Socratic approach may not require the coach to have any safety experience at all. Coaching proper is industry or profession-agnostic when done professionally. The coaching (Socratic based) is centred around the individual and their internal motivation, biases, desires, fears, etc. I've personally coached where I didn't have a clue about the person's role, or the industry they worked in! That being said, my experience is that when we are sent into an environment as a safety coach, there is an expectation from everyone that the safety coach will have some safety skills and experience. Whether it's actually needed or not is another matter.

In Lloyd's (2020) book *Next Generation Safety Leadership: From Compliance to Care*, he breaks down trust into three components: Ability, Benevolence, and Integrity. Although the Integrity aspect is the most important part of facilitating and maintaining trust, the Ability aspect is also important. Imagine the following scenario.

Client: We want to engage safety coaches to work with our leaders and help them build a resilient and capacity-focused organisation.
Consultancy firm: Great! We have lots of coaches who would be great! How soon do you need them?
Client: Coaches? Do you mean safety coaches?
Consultancy firm: Sort of. They are highly skilled organisational coaches.
Client: But do they have any experience in safety?
Consultancy firm: Well, no. But you don't need safety experience to engage with individuals in a coaching relationship!
Client: Yeah, but … this is all about safety improvement!

See how the Ability aspect comes into question? I 100% agree that organisational coaches don't need specific technical knowledge when coaching clients from backgrounds and industries the coach is not familiar with; however, when the context is specific, and the goal is defined to a particular parameter, i.e., safety performance, it is my opinion that Socratic-based coaching benefits from the coach having at least some safety experience, especially the more holistic organisational perspective safety experience. That being said, I'm sure there are many professional coaches out there who greatly benefit safety initiatives, regardless of their "safety experience." It just sounds like a harder "sell" to me!

EMPOWERED SOLUTIONING

And this, my friends, leads us to what Empowered Solutioning is: **A safety improvement approach which utilises aspects of safety knowledge and coaching principles, to varying degrees.** The capability to transition through and between the aforementioned "types" of safety coaching identified so far.

Here's a breakdown of the term Empowered Solutioning to reinforce what I've laid out above.

"Empowered" reflects the coaching aspect, where clients are encouraged and enabled to develop their own solutions and take ownership of their decisions through self-reflection, challenging internal biases and axioms, goal setting, and being held accountable for their continued development.

"Solutioning" is a term often used in consulting (for this updated version, we can view this as "having specific contextual knowledge") to describe the process of developing solutions to organisational and systems challenges. The safety and technical knowledge required for quick and measurable interventions, as well as an organisational and holistic approach more generally (dependent on the skill sets held by the practitioner and the safety coaching "type" that's required).

I've posted about Empowered Solutioning on my website in a post. But in case you haven't jumped onto my website yet (why haven't you?), here are the benefits as I see them for safety practitioners to adopt an approach such as Empowered Solutioning.

Leverages Expert Knowledge

Depending on the requirements of the project, expert knowledge can be utilised, be it highly technical knowledge of specific systems or environments, such as specific knowledge of high-hazard environments like mine sites, exploration (oil and gas), or manufacturing and construction, as well as detailed technical knowledge, such as technical standards or even lived experience with particular systems, tools, or processes.

Encourages Active Participation

The coaching element, which by default is more of a "pull" conversation (questions, enquiring, curiosity), enables safety practitioners and leaders to build trust, empower individuals, and promote engagement. Using Table 3.1, we can see that this varies depending on the role of the safety coach at the time, but it could be said that there are elements of this across all three approaches, especially the Socratic and Analytical approaches.

Facilitates Co-created Solutions

Again, dependent on the approach being used at the time and in comparison to a pure compliance and control approach to safety, Socratic and Analytical both co-create, to a greater or lesser degree. Co-creation of solutions isn't just a better way to come up with better solutions; it's also a better way of getting buy-in to change!

(Scan the QR and search my YouTube videos for "Humanising safety")

So the bigger question is, why bother giving this approach a name? Isn't that just a bit vain? Maybe. But as I've addressed constantly throughout this book, safety coaching has a definitional problem. What it is, what it does, and how. Even coaching proper can be difficult to define and describe, as I discussed with the fabulous Charity Bekker in a YouTube discussion. So, adding the complexity of "safety" into the definitional mix only complicates things even more! So, it's my hope that as safety coaching becomes more and more mainstream, which I have no doubt that it will, people will find the use of the term Empowered Solutioning easy enough to remember, and descriptive enough to explain what a safety coach does.

It is also my hope that one day I'll see posts or company statements with sentences like "As part of our safety leadership we use the Empowered Solutioning approach." Or even (heaven forbid!) job roles with titles such as "Empowered Solutioning Practitioner." Now, wouldn't that be something!

HOW DOES ALL THIS TIE INTO PART II?

Throughout Part I, there are a few references to how leaders can utilise some of the skills of coaching as a way of developing their own people. And in fact, there is a whole research literature on coaching as a leadership style. To this point, I will be looking at how coaching skills can help facilitate "safety leadership" in particular.

So, what can you expect in Part II?

A look at what safety leadership is, an overview of what a coaching leadership style is, and then a look at how approaches, such as New View Safety, can be better facilitated when leaders adopt a coaching leadership style.

The obvious question here (if you haven't already considered it) would be, why would I want to include these in this book? Surely I want contracted independent safety coaches to be the arbiters of all knowledge and specialised skills to assist companies with their transition or improvements? Well, no, actually, I don't! This isn't about me, or the development of safety coaching (well, maybe a little!); it's much bigger than that. It's about safety improvement. Period.

I learned a lot from my time as a consultant. One of the key aspects I took away was the philosophy of: we don't want to be one of those consultancies that hang around forever, constantly charging the client for a service that they themselves could learn to apply without us being there.

Maybe I'm not capitalistic enough. Maybe I'm naive. Maybe I'm too idealistic.

Maybe.

But this philosophy resonates with me.

So, for me, the days of consultancy firms holding onto knowledge and ideas that could potentially benefit more organisations (as they do via IP, trademark rights and sky-high pricing) aren't how I want to operate. It's had its day, and there are plenty of cautious and wary people out there to verify this! Yes, people need to learn new skills and potentially adopt new ways of operating and leading, but this doesn't mean there needs to be an embedded "specialist" delivering trademarked content to the select few companies who can afford it! As I've already mentioned, this may seem uncapitalistic, naive, and idealistic, especially considering the fact that I am one of the co-founders of a training company which specialises in safety coach training (minor plug here for Safety Coach International); however, our core principle is to train, prepare, and encourage participants to quickly and effectively understand and apply the skills associated with safety coaching in complex environments. Not to embed ourselves within an organisation over years and years, delivering trademarked content, and all the while charging exorbitant hourly fees! There needs to be self-sustaining and realistic solutions if we are ever going to move the dial on safety improvements via safety coaching and safety leadership!

QUESTIONS TO POKE YOUR BRAIN!

- Question 1: When an incident occurs, how might the natural human tendency towards "Hindsight Bias" be colouring your perception of what the worker "should have known" in the heat of the moment?
- Question 2: How might shifting your conversation style from "compliance" to "curiosity" help you bridge the gap between Work-as-Imagined in your procedures and the messy reality of Work-as-Done?

Introducing the Integral Model 4

INTRODUCING THE INTEGRAL MODEL

Part I was framed through an "individual" lens. The focus was on how individuals' internal biases, beliefs, and assumptions can help or hinder the application of New View Safety. This part of the book needs more of a holistic lens through which safety leadership can be framed. Yes, we need to factor in individuals' perceptions and beliefs as these are important; however, we also need to zoom out from the individual conversations to a broader map of leadership demands that incorporates organisational culture and systems. This systemic meta-framework is not just useful for leaders in general; it is also very useful when we discuss the approach needed for New View Safety, which, as we know, requires more of a systemic and holistic appreciation.

For this meta-framework or lens, I will be using Ken Wilber's Integral Model, which is useful on many levels of analysis, from individual beliefs to organisation-wide systems and processes. But before I continue, let me clarify a few things. I'm proposing the Integral Model as a practical lens only, not a "total truth" solution. I'm using it as a philosophical and integrative framework to raise awareness and curiosity, not as an empirically validated model of unquestionable truth. I'm not using the full AQAL approach either; I'm restricting use to the four quadrants to keep it usable in the context of this book. That being said, it may be useful for you to note that the Integral Model is used within leadership and coaching practices around the world as a framework for working with organisational complexity and for making explicit the interconnected psychological, relational, cultural, and systemic dimensions of organisations. Think of it as a model to counter siloed or fragmented thinking. It helps shift our thoughts from purely "behaviours" or "processes" to a broader appreciation of other factors that contribute to the overall performance of an organisation. The parallels with New View Safety are striking. They

 DOI: 10.1201/9781003590538-5

both seek to broaden myopic attention from any one particular area of focus (behaviour, systems, etc.) to a broader, more holistic interpretation of all the facets of an organisation.

So, for those of you who are just being introduced to the Integral Model for the first time, here's what it is in simple terms, in the context of safety leadership.

Safety leadership that factors in the interactions between four quadrants: I (psychological), We (social), Behaviours (actions), and Systems (tools).

I – *Inner Experience* Thoughts, feelings, beliefs, values, and presuppositions of an individual. Examples: A supervisor's belief that "Speaking up will get you in trouble!"; a frontline worker's anxiety after a near miss; a leader's personal commitment to learning rather than blaming. (Known as **UL** for **U**pper **L**eft Quadrant)	Behaviours – *WhatPeopleDo* Observable actions, skills, habits, and results. Examples: A leader stops work when conditions change; workers taking short-cuts; supervisors holding regular learning-team style conversations. (Known as **UR** for **U**pper **R**ight Quadrant)
We – *Culture and Relationships* *Shared meanings, norms, stories, and levels of psychosocial safety.* Examples: "Around here, we don't report small errors!"; an unwritten rule that production always beats safety. (Known as **LL** for **L**ower **L**eft Quadrant)	Systems – *Structures and Environment* *Formal processes, technology, resourcing, rules, KPIs, and work design.* Examples: Incident investigations processes focused on "blame and punishment"; staffing levels and fatigue policies; design of permit-to-work systems; how safety performance is measured and rewarded. (Known as **LR** for **L**ower **R**ight Quadrant)

I haven't included the many arrows pointing from each quadrant to every other, and back, for ease and simplicity of viewing the four quadrants, but the four quadrants are not siloed or standalone; they are all interconnected or co-nascent (born together) as Wilber describes them. They co-evolve and inform each other. Let me demonstrate with an example.

Root Cause Analysis

Consider investigating a major industrial accident, such as a chemical spillage or equipment explosion.

UR: Safety investigations usually start here, with the observable actions, "What did the operator do, or not do?"

LR: And then the results, usually once a conclusion about blame has been formed, "What mechanical failure occurred?", "What chemical reaction took place?", "What were the effects or potential effects?"

Using New View Safety principles, as well as the Integral Model, we know that siloed thinking only gives us one part of the story. Let's look at how the other quadrants might influence the behaviours and actions that came to be.

UL: What internal beliefs did the operators have? *Were they afraid to report warning signs? Were there near misses that could have prevented the incident if the workers felt safe to report them?*

LL: What cultural or relationship aspects could have facilitated this incident? *Is it an unwritten rule that production is king, regardless of safety? Is reporting hazards seen as a waste of time and useless? Does the organisation view accidents as reasons to blame and punish, or opportunities to learn and improve?*

LR: What structures and environmental aspects contributed? *Was the maintenance budget cut? Are the isolation permits fit for purpose, or too complicated and difficult to obtain? Is there proper system integration across multiple platforms to support effective monitoring?*

This is a useful, although basic, demonstration of what is meant by "holistic" or "systemic" view of safety. By referring to the Integral Model and recognising where we are focusing (in the example, it was on behaviours and outcomes), we can learn to be more curious about how the other quadrants might have influenced the outcome or behaviour (**UR**). As we work through the rest of the book, the Integral Model will help scaffold what we discuss, and you will notice that the focus of coaching leadership is on three of the quadrants (**UL**, **UR** and **LL**): internal beliefs and views, behaviours and results, culture and relationships. This isn't to say that the structures and the environment quadrant (**LR**) isn't valid; it definitely is. However, the benefits brought about by coaching leadership are in understanding the catalyst for observable actions and behaviours through understanding internal motivations and cultural norms and how they interact with the "current" processes and systems; systems and processes can then be modified to support desired behaviours and beliefs if needed. This is how we begin to gain a more holistic and systemic understanding of the emergent property of safety within our organisations. Although, as we have already noted, it's harder to achieve than merely fixing what we can see! The old saying of "if you can't measure it, you can't manage it" comes to mind here, and although we know this isn't true, the heuristic still prevails within our cultures and organisations.

The purpose of introducing the Integral Model isn't to generate discussion about the model itself; rather, it's to provide a recognised and simple map to help generate better and more curious safety questions. It can help leaders move beyond one-dimensional explanations (he did this, she didn't do that) and instead ask questions that connect people, culture, behaviours, and systems. Try to see it as a superhero, the "Destroyer of Siloes." I'll be referring back to our superhero as we make our way through the rest of this part of the book.

If you are interested in how the Integral Model fits with safety performance, and organisational performance more broadly, and want to learn more, I highly recommend you read *Safety Performance Reimagined: A 4D Approach to Organizational Performance* by Read and Ritchie (2021). In their book, Brett and Rod draw on the Integral Model theory as one of the foundations for their 4D Safety Framework, using it to integrate the human, technical, organisational and market dimensions of safety performance into a single socio-technical view.

Try This

Take a recent near-miss or incident and write one short bullet point under each quadrant: what people were thinking, and why (**UL**), what they did or didn't do, and why (**UR**), what the team/organisation norm is, and why (**LL**), and what system conditions were present, or not present (**LR**).

LEADERSHIP STYLES AND NEW VIEW SAFETY

I can wholeheartedly state that the field of leadership is still somewhat of a mystery to me! There are too many "styles," "approaches," or "techniques" for me to get my head around; from Autocratic to Democratic; Transformational to Transactional; and let's not forget Servant and Eco leadership! All of them (and many more I haven't mentioned!) have their pros and cons, benefits and pay-offs. As with most complex things, context is king, and my takeaway from having to write about a lot of these approaches and techniques is:

1. None of them seems to account for or factor in innate personality traits, or historical experience and characteristics of the individual. They all seem (from memory, although I'm loath to go back over

my lecture notes to verify this!) to have a list of behaviours, desired outcomes, or structured thought processes required for the performance of the particular style or approach.
2. They are all context-specific, to a greater or lesser degree. By that, I mean they are dependent on the environment and desired outcomes. Would, for example, democratic leadership work well on the frontline of a battlefield, or for emergency services being called to a catastrophe? Likely not. However, if you work in an environment that requires buy-in, such as product development or community engagement, this style would be right at home and work a treat!

It is for these reasons that I truly believe that the time and effort put into fully mapping out our journey (or terrain) of how we have come to be the leader we currently are is time well spent (more on this in the next section). Other tools and techniques can also facilitate deeper awareness of self, such as Dominance, Influence, Steadiness, Compliance/Conscientiousness (DISC), which is a descriptive behavioural-style assessment and training tool that broadly aligns with the Big 5 personality traits there are also dedicated Big 5 personality trait tools you can use; these can give you a more detailed picture of your inherent personality traits. If we overlay the Integral Model here, we can see that both DISC and the Big 5 can be used as indicators of how we perceive (**UL**) and how we act (**UR**). They say very little about the cultural (**LL**) or systems and processes (**LR**) quadrants, although, as we will see, **LR** aspects can significantly influence what shows up in the **UL** and **UR** quadrants. A word of caution here: all the tools mentioned should be viewed as pieces of a puzzle, and not an absolute final diagnosis of oneself.

So, which leadership styles or approaches are best for facilitating New View Safety? I think this is easiest to discuss when we look at the styles and approaches that we know aren't conducive to New View Safety principles, then apply the opposite. Yes, you could say this is an a priori way of framing the question, and you'd have a point. However, in the spirit of Occam's Razor (the simplest explanation that fits the facts), starting with what obviously doesn't support New View Safety and inverting them is a reasonable place to begin. Another cautionary note to mention: don't read the following and imply an "all or nothing" lens, i.e., we know that "asking" as opposed to "telling" is preferable, but there will be times (context dependent) when telling is absolutely needed. Although they may be written as absolutes, they infer a level of granularity, especially if you're judging yourself against them.

LEADERSHIP STYLES AND APPROACHES THAT WON'T FACILITATE NEW VIEW SAFETY	OPPOSING LEADERSHIP STYLES AND APPROACHES
Autocratic Leadership: emphasises obedience and control from a singular source with little input from workers (UL). Rigid hierarchies and decisions concentrated at the top (LR).	*Participative or Democratic Leadership:* openness and humility (UL). Participative structures and joint decisions (LR).
Command-and-Control Leadership: belief that leaders must direct and workers must comply (UL). Detailed procedures are the primary control (LR).	*Enabling or Empowering Leadership:* trust in frontline judgement (UL). Systems that decentralise authority where applicable, and support local adaptation (LR).
Punitive or Blame-Focused Leadership: attribution of problems to "bad" individuals; low empathy (UL). Disciplinary policies, zero-tolerance rules, and a blame focus in investigations (LR).	*Just Culture and Learning Oriented Leadership:* curiosity and fairness (UL). Learning reviews. Clear and appropriate accountability frameworks (LR).
Transactional Leadership: focuses on control via incentives and sanctions (UL). Bonus schemes and behavioural scorecards as measurement mechanisms (LR).	*Transformational Leadership and Coaching-Oriented Leadership:* developmental, purpose-driven focus (UL). Coach training and routines, reflection practices, and developmental pathways (LR).

The table shows how the individual leaders' values and beliefs (**UL**) are integrally connected to the organisational systems and processes (**LR**). Let me demonstrate.

What Would Be the Likely Outcome of This Scenario?

Leader 1: has a natural disposition to curiosity, building trust, and fairness (**UL**). However…

Organisational Systems 1: promote and enforce transactional and autocratic behaviours from the leaders (**LR**).

Likely, the leader will experience cognitive dissonance. What is being asked of them won't align with their sense of "self" or "purpose." This is what is meant when people say "find a company that aligns with your values."

What About This Scenario?

Leader 2: has a natural disposition for command and control. They believe leaders direct and workers comply, and hierarchy must not be challenged (**UL**).

And…

Organisational Systems 2: promote and enforce transactional and autocratic behaviours from the leaders (**LR**).

The leader would be happier than Leader #1 for sure because there is more alignment between the leader's values/beliefs (**UL**) and supporting organisational systems/processes (**LR**). As the systems better align with the leader's intrinsic style and traits, the ability of the leader to enact the systems and processes becomes more proficient than in scenario #1. This could also be said for leader #1, if the systems and processes (**LR**) were reversed.

Here is a good moment to link to a HOP principle: context drives behaviour. If you want different behaviours from your leaders or anyone, the systems and processes (context) must promote and support the desired behaviour. Purely asking leader #1 to be more autocratic or leader #2 to be more collaborative won't work. They will sit through classes and presentations, maybe even agree in principle; however, when it comes to the daily grind and pressures that come with it, they will default back to what comes intuitively: their natural traits and values.

Using the table, we can see common threads or concepts throughout the styles and approaches that are opposite to ways we know won't facilitate New View Safety principles. These can be categorised as "learning orientated," "relational," and "participative." These are important to identify as they clearly show that in order for New View Safety to be enacted in an organisation, the leaders can't use styles or approaches that align more with aspects of command and control or autocratic, as they won't work.

However, before we get to the nitty-gritty of coaching style leadership, I'd like to reinforce the importance of self-awareness for leaders. Throughout this book, I've made constant reference to the importance of individuals starting work "within" first, before any work can start with "others."

Try This

Identify one organisational system or process in your workplace that you believe is actively working against the leadership behaviours you want to demonstrate. What would need to change, and who has the authority to chan

---

LOCATING SELF AS A LEADER

For me (and you, I presume, as you're reading this book!) I have a great interest in coaching, which focuses on self-reflection, awareness, and clarity of desired states. So imagine my joy when I was studying for my post-grad in Organisational Coaching and Leadership, and I came across the work of Simon Western! Who's Simon Western, I hear you ask.

Dr Simon Western is a leadership and coaching practitioner and a well-respected author of several books; he's also a super interesting guy if you're interested in doing your own research on him. In his book *Coaching and Mentoring: A Critical Text* (Western, 2012), he outlines an exercise he uses to help leaders; it's called "Locating Ourselves." Think of this as a "meta" approach to leadership improvement, if you will; the thing that comes before you decide which approach will best suit your particular environment and requirements. As with coaching, a clear understanding of the terrain (reality) is required before we can make efficient progress in any particular direction. And that's what Western has developed with great success.

Without the terrain being properly mapped out, and a clear understanding of "how" we got to our particular place (current type of leadership), how on earth can we effectively and efficiently plot a path forward to a desired future leadership type? Like with maps, we need to know where we are located before we can proceed with any accuracy; with Simon's "locating ourselves" he's effectively saying we also need to know how the journey "so far" has influenced us, for we are not just a moment in time; rather, we are a collection of our innate capabilities (and inadequacies) and past experiences.

Here is an extract from an essay I wrote back in 2023, if the above didn't land with you:

> Much like the factors that determine how a tree is represented in the world, leadership is shaped by a combination of factors, including our genetics, immediate environment, and context. Our genetics, or innate qualities, such as our personality and temperament, play a role in how we lead. Our immediate environment, or the people and experiences that surround us, also influences our leadership style.

Simon's "locating ourselves" is a discovery for the individual. A mapping of the biography, relationships and contexts that have shaped them into the current leaders they are. Once leaders are aware of "how" they became the leaders they are, they can be more aligned with their values and authenticity, as well as fully appreciate what it would take for them to modify their leadership style, should they wish to.

Note: We will come back to this self-awareness perspective later in the book through a different author, Edgar Schein.

See why I wanted to discuss Simon?

Coaching is a map for the terrain going forward. We map opportunities, hurdles, and the effects of unconscious patterns (biases, axioms, etc.) on desired future states.

Simons' "locating ourselves" is a map of the terrain of "how" we got to where we currently are, and how that journey made us into who we are, with our unconscious patterns and inherited norms that we collected along the way.

> You must become master of yourself, ruler of your own mind; for it is not that we have a short time to live, but that we waste much of it.
>
> Seneca

Put these together, and individuals have a far greater chance of appreciating what type of leader they are, why, and where they want to be. Overlaying the Integral Model again, we can say that it's a discovery of the "I" (**UL**); not just that we have internal values, beliefs, and principles about leadership, but also where they came from. The map* (past, present, and future) is clearer and better defined, and one can navigate in the "now" better, as well as plan a realistic potential future state from a well-defined point, which is critical for navigating effectively.

*Using the metaphor of maps is a well-used psychological approach, as it fits nicely with how we actually perceive the world. For example, we don't perceive everything around us all the time; there are just too many things we could pay attention to. Instead, we focus on and give our attention to things (the terrain of life) that either enable or impede our progression towards a desired state or outcome. As we travel through life in a desired direction, the terrain is perceived by us accordingly; we look for things to enable progress and things that could impede progress. Each time we perceive either an impediment or an enabler, our body reacts according to our perceptions, i.e., as we encounter things that enable us to achieve what we would like, we get a hit of dopamine, which not only feels good, but also reinforces motivation and behaviours. Conversely, when we encounter things that impede our progress, our body produces the neurotransmitter norepinephrine and a steroid hormone, cortisol. This combination forms part of the fight or flight response, and we can become reactive to perceived threats. This is how our internal chemistry tracks our journey through life, and rewards or prepares us accordingly. The driving example in Part I where we are late for a meeting at work, comes to mind here as we think about our emotions. Things don't just cognitively feel good or bad; we are biochemically prompted to react a certain way, in a bidirectional relationship. If we therefore understand, through deliberate self-awareness, the perceptions and beliefs that shape how our body reacts to situations, we can begin to modify how we perceive and, therefore, how we react. It's this redefinition of perceived threats and goals that also makes New View Safety potentially difficult to come to grips with. If we view any incidents and errors as impediments to a desired goal (low LTI, zero harm, etc.), we will biologically react as if it's a threat, which will shape how we relate to the situation. If we perceive incidents or errors as opportunities to learn and improve, we are far less likely to become reactive and can actively engage curiosity as a response mechanism.

Try This

After a safety incident, jot down: "What part of my personal story is driving how I'm reacting to this right now?" Then write whatever comes to mind. Don't filter or sanitise your thoughts; they are only for you to see and think about.

PART II

Leader as Coach 5

LEADER AS COACH

Focusing as we are on three quadrants of the Integral Model (**UL**, **UR**, and **LL**): internal beliefs and views, behaviours and results, culture and relationships; I want to begin with the actions of the leader (**UR**) and look at how a coaching style of leadership can be enacted. Before we proceed, though, let me lay out a definition of coaching leadership as a way of defining exactly what I am speaking about.

A coaching leadership style is easiest to define if we compare it to its extreme opposite. This is useful as a definitional aid, but it isn't useful as a method of classification for individuals such as ourselves, as there is a huge array of "greyness" in between the two definitions, which sit on opposite ends of a continuum. With that said, here is a juxtaposed definition of coaching leadership. A behavioural style of leadership that's less directive (command and control) by default, and more facilitative and curious (Socratic and active listening) and focuses on support, guidance, and empowerment.

Now, if you're anything like me, you might read that description and think something like "OK, sounds interesting. But what does that look like in practice?"

Well, in order to answer this question, I'm first going to introduce you to a phenomenal man called Michael Bungay-Stanier, or MBS as he goes by. If you are looking to learn about coaching leadership style, you can't go wrong starting with him. Bungay-Stanier (2016) is an international bestselling author whose book *The Coaching Habit: Say Less, Ask More & Change the Way You Lead Forever* is widely regarded as one of the most influential coaching books of the century. He's not just an author, he's also a recipient of the Rhodes Scholarship and has degrees from the Australian National University (ANU) and Oxford University. His career is impressive, with experience in consulting, branding, and entrepreneurship across multiple countries. And to top it off, he was also named "Canadian Coach of the Year" in 2006.

DOI: 10.1201/9781003590538-7

Without going into great detail on how MBS helps operationalise coaching leadership, below are some takeaways from some of his books. Though I would highly recommend you read his books for yourself, as they are not just transformative, they are also very easy to read and connect with.

MBS outlines three core principles to help people become more coach-like in their everyday leadership. These two may surprise you!

Be Lazy! By this, he is referring to the innate desire some leaders have to provide solutions to problems. This not only makes you the source of all solutions for your team, thereby unintentionally disempowering them, but it also adds to your existing workload! Being lazy in the sense laid out by MBS isn't a pejorative term; it's meant as an enabling term, for both the team you lead and for yourself. If your team are "trained" by you as their leader to be creative in their ability to solve problems, without you, just imagine how many problems they could solve, all while you spend time on what's most important to you and reap the glory of a high-performing team.

Be Curious. Being curious is a particularly important principle for leaders who like to rush to advice, the Advice Monster as Bungay-Stanier (2020) calls it in his book *The Advice Trap: Be Humble, Stay Curious & Change the Way You Lead Forever*, and is addressed along with the above principle. Even when we need to provide advice, if we stay curious that little bit longer, we can gain a better understanding of the actual problem, as opposed to assuming we understand all the nuances immediately. This reflection time of curiosity used by the leader may even lead the questioner to come to their own self-discovery of the potential solution, thereby facilitating the principle of Be Lazy!

MBS operationalises the coaching leadership approach through Conversation Flows and Seven Essential Questions that are easy to learn and put into action day-to-day. I'm not going to discuss these here, but I will say that his questions are broadly aligned with the format of the GROW methodology, which, as you already know, is a widely used framework in evidence-based coaching.

Try This

For the next week, commit to responding to every problem brought to you with two genuinely curious questions rather than one answer. Notice what happens to the conversation, both to the person and to yourself.

Let's now look at the organisational cultural benefits (**LL**) of using a coaching style of leadership, because as we know, all the quadrants are interrelated and influence each other, and as we will see, the leader's actions (**UR**) are intrinsically linked to the culture (**LL**). Imagine what could happen if, as a leader demonstrating genuine curiosity and a learning behaviour, your entire team has the confidence and capability (read: feels safe) to do the same. What would be the effects of this? This is exactly what Edmondson (2018) explored in her book *The Fearless Organization: Creating Psychological Safety in the Workplace for Learning, Innovation, and Growth.* Edmondson addresses psychological safety and how this is one of the most critical enablers of high-performing teams. These teams reported more errors, not fewer (observed in her hospital research), as they felt safe to admit mistakes and had a desire to learn from them as opposed to blaming individuals involved. They also had increased creativity, adaptability, and resilience. Edmondson highlights, though, that psychological safety within a team is not self-generating; it is largely a product of the leader's behaviours. Behaviours such as inviting input, responding non-defensively to bad news, and modelling curiosity (**UR**), all aspects that we've touched on. Edmondson's insights are not just academic. In 2012, Google began an internal research project called Project Aristotle, with results released years later. The project studied hundreds of Google's own teams to identify the special "ingredient" of high-performing teams. Their presupposition was that the "who" of the team, i.e., who was in the team, and the skills they came with, would be the special ingredient. What they ended up discovering was that the "how" of the team mattered more, i.e., how team members interact and communicate with each other (psychological safety), proving to be the greater predictor. Drawing on both Edmondson's research and Google's findings, we can reliably state that, if you want to create the culture for a high-performing team (**LL**), and all the benefits that come with it, then it begins with the leader creating psychological safety through their actions (**UR**), which can then be emulated across the team.

Try This

Consider opening your next team meeting with this simple, genuine question: "What's one thing that's been on your mind lately that you haven't felt comfortable raising?" Notice what comes up (if nothing is volunteered, this could also be a good indicator for you), and notice your reaction to it.

So, what do we know so far? We know that a coaching leadership style (**UR**) is complementary to enacting New View Safety principles and strongly associated with, and a catalyst for, psychological safety within teams (**LL**). So, the obvious next question is, if we know all this, what shapes the leader's leadership style? If the benefits are obvious, then leaders can just change their leadership approach, right? Well, no, actually. As we've already demonstrated, and we'll touch on again shortly, individuals' perceptions, values, and beliefs shape their actions. Skills or knowledge are essentially filtered through these perceptions, values, and beliefs, and only the skills or knowledge that align with these will become routine behaviour.

One potential answer for where we get some of our perceptions about leadership could come from Schein's (2013) book, *Humble Inquiry: The Gentle Art of Asking Instead of Telling.* Schein invites us to go deeper, as with Simon Western's "locating ourselves," and to look at the very environment (culture) as a source of influence for our leadership presuppositions and deeply held beliefs (**UL**). Schein puts forward the idea that many Western cultures have conditioned us, over time and through cultural artefacts, to believe that competent leaders have all the answers, and that they lead by command and control; and that it is this cultural "norm" which has shaped us into organisational leaders who "tell," as opposed to leaders who can "ask." Humble Inquiry is Schein's proposed practice to help leaders shift this internal perception from a "tell first" default to a genuine curiosity-driven approach, one that benefits not just the leader, but the individuals they are responsible for and the organisation as a whole. If the question "but what does genuine asking over telling actually look like?" is in your head right now, allow me to give a brief overview of Schein's book, although I highly recommend reading it yourself for a more detailed understanding. Schein highlights that there are many types of questions, and they are not all "humble inquiry." Diagnostic, confrontational, or rhetorical questions don't qualify because their intent isn't driven by genuine curiosity to learn something you don't already know. Intent and relational interaction are at the heart of Humble Inquiry. If you're only asking questions to catch someone out, put them under pressure, or confirm what you already believe or know, then they can be called lots of things; Humble Inquiry, as Schein proposes, isn't one of them. Schein also argues that Humble Inquiry builds trust between people, thereby allowing honesty and open communication, all vital aspects for a psychologically safe team, especially when looking to adopt New View Safety principles.

Try This

For the next two weeks, keep a simple tally. Every time you respond to a situation by telling, mark down a T. Every time you respond by asking, mark down an A. At the end of the two weeks, take a moment to reflect on your ratio. What does it tell you about your default leadership style?

What I hope to have highlighted in this section is the logic of the Integral Model in relation to a causal chain across three of the quadrants. With Schein, we have assumptions and beliefs of the individual (**UL**), which can shape and influence the actions and behaviours we might want to enact, such as those highlighted by MBS (**UR**). It is these actions and behaviours which, in turn, create or hinder cultural conditions that enable team performance and safety, as discussed with Edmondson (**LL**). Although none of these authors is explicitly writing about safety performance per se, they do provide reasoning for facets that we know contribute greatly to New View Safety leadership. Free-flowing information, psychological safety around discussing and learning from mistakes, and techniques that we can begin to incorporate. The next logical question, then, is whether there is evidence for a coaching style of leadership? And, if so, what does it say?

EVIDENCE FOR COACHING LEADERSHIP

We've covered what a coaching style of leadership is and how the Integral Model shows the conceptual interconnected nature of it, not just from an organisational perspective, but also on an individual level for the leader. In this section, we are going to look at what the empirical evidence says about this particular style of leadership on aspects such as performance and trust. Two peer-reviewed articles form the basis for what I will cover; below, there is a brief overview of each, as well as being provided at the back of the book for your reference.

Coaching Style Leadership

Kunos (2018) explores how managers can act as agents of development by using coaching tools on a daily basis to help their employees learn and improve performance in the "Coaching Style Leadership" (*International Journal of Social Science and Business*).

Coaching Leadership and Creative Performance

Wang et al. (2023) demonstrate in their article "Coaching Style Leadership and Creative Performance" (*Frontiers in Psychology*) that when leaders act as coaches who provide direction, mental support, and feedback, they significantly boost employees' psychological empowerment and creative performance.

An important caveat here is that none of the research we will look at talks about a coaching leadership style in the context of New View Safety, or even safety generally. We will, however, align the researched benefits with New View Safety and demonstrate how, as with coaching proper, the benefits support the principles of New View Safety.

Across multiple studies, a coaching style of leadership is linked with higher levels of creativity, a stronger culture of learning, and improved performance. There are several mechanisms of a coaching leadership style that facilitate these benefits, with psychological empowerment playing a central role and fostering the trust that is vital for people to report news to the leader, especially if it's bad news!

Improved performance is a broad term. In relation to New View Safety, "performance" is not just output or compliance; it's the capacity for people to care about (**UL**), talk about (**LL**), and ultimately act upon real-world safety problems. Research shows that a coaching leadership style builds these capacities by strengthening people's sense of agency and voice, which in turn drives the psychological safety required for creative problem-solving.

Now, whenever I drop the "psychological safety" line, the responses can be varied, depending on the audience. Some people get the distinction between psychosocial risks and psychological harm; essentially, psychosocial risks (workload, role expectations, etc.) can lead to psychological harm (stress, anxiety, burnout, etc.). Having low psychosocial risk with elevated psychological safety (feeling safe to take interpersonal risks), we have the prime conditions for creativity, problem-solving, and constructive communications.

Others, however, have a different response. They perceive psychological safety as "being nice" or "letting people off the hook." If this latter perception resonates with you more, that's fine, I'm not judging you. Let me share the following with you, though, as a way to potentially spark some curiosity about your beliefs. Studies show that when coaching proper (not leadership coaching) is utilised post-training, the retention and transfer rate is higher, quite significantly higher (5–10% to approx. 90%). Now, here's the link I'm making for a coaching leadership style. If post-training coaching is proven to be so much more effective in helping retention and transfer for embedding new skills and approaches, a coaching leadership style that utilises coaching techniques, including creating the psychological safety aspect, is likely to support learning retention and transfer performance more than a purely directive style, as the approach reinforces the conditions on a daily basis. Psychological safety is no longer seen as "being nice" or "letting people off the hook," it's a must-have for high-performing teams, as demonstrated by Google's Project Aristotle.

A coaching style of leadership gives leaders concrete behaviours that build trust, while still holding people to account. Trust is essential if we want workers to disclose uncertainty, near misses, systems and personal failures; all of which are paramount to the enactment of New View Safety principles.

COACHING LEADERSHIP THROUGH THE INTEGRAL LENS

Part I covered professional coaching as a 1-on-1 application; this is what you could call the "classic" notion of professional Organisational Coaching. Here in Part II, we are looking at how some of the skills of professional coaches can be utilised by leaders as a "style" of leadership. It's not lost on me, though, that some of you will still be carrying this preconception about coaching being "1-on-1" into this part of the book, so I thought now might be a good time to broaden our perspectives again about coaching by using the Integral Model.

Let me start with this. A coaching style of leadership isn't just about better 1-on-1s; it's not just about increasing another individual's safety performance; it's much bigger, as you'll see in the table below, which has the four quadrants of the Integral Model, plus some examples that a coaching style of leadership can bring to safety (and other performance metrics) across an organisation. When you're reading them, ask yourself, is this the kind of leader I would want to report to? If your answer is yes or no, ask yourself why.

I – Inner Experience Thoughts, feelings, beliefs, values, and presuppositions of an individual. Sees workers as resourceful experts, truly believing that "people adapt to make things work; my job is to understand those adaptations, not to fix the people." When things go wrong, reframes "they were careless" to "given the conditions, how did this make sense?" Becomes genuinely curious about the complexities of the work done. Connecting with workers through meaning "I'm here to learn how work is really done so we can improve, not to catch you out."	**Behaviours – WhatPeopleDo** Observable actions, skills, habits, and results. Opens conversations with "Walk me through how this usually goes on a normal day" or "What made today different?" Asks systemic questions out of curiosity, such as "What were you balancing here?" and "What pressures were in play?" and not just on individual choices. Uses language that separates people from problems, such as "Where did the work setup make it hard to succeed today?" and "Which system or process causes the biggest frustrations?"
We – Culture and Relationships *Shared meanings, norms, stories, and levels of psychosocial safety.* Builds shared stories about "work as imagined" vs "work as done" and promotes curiosity about the gap. Promotes "success under pressure" stories, so learning about what goes well becomes as legitimate as incident investigations. Reinforces norms that speaking up about weak signals, near misses, and workarounds is an act of professionalism and care, not disloyalty.	**Systems – Structures and Environment** *Formal processes, technology, resourcing, rules, KPIs, and work design.* Uses learning reviews and learning teams to explicitly ask about conditions, adjustments, and system performance; not just violations. Aligns KPIs and dashboards to include indicators of learning instead of only injury rates or rule violations. Builds processes that value worker participation in risk assessments, procedure design, and change management, thereby making "workers as problem solvers" a reality.

As demonstrated above, a coaching style of leadership actively affects every facet of an organisation. As a leadership style, it creates the space, safety, and support needed to gain trust, which benefits everyone and is critical for leading an organisation into New View Safety.

Try This

After a conversation (with an individual or a group), ask yourself: "What shifted in the person/s thinking (**UL**), behaviour (**UR**), the team norms (**LL**), or how we'll work the systems (**LR**)?"

RAISING THE BAR ON SAFETY USING A SIMPLE TOOL

How many acronyms do you have in your profession? If you work in safety, I already know you have a plethora to remember. Just looking at ways to classify incidents, we already have Lost Time Injury (LTI), Medical Treatment Injury (MTI), Total recordable Injury frequency Rate (TRIFR), Significant Injury (SIFR), and First-Aid Injury (FAI) to name a few! Safe Work Method Statement (SWMS), Job Safety Analysis (JSA), As Low AS Reasonably Practicable (ALARP), and Personal Protective Equipment (PPE) form the majority of conversations we have on a daily basis in the safety world.

Acronyms can be useful; they can also be overwhelming and confusing, especially to people who don't work within the profession for which they were created.

I've worked in several other industries; from IT to finance, agriculture to theatre. They all had their own special acronyms. My personal favourite was when I worked in IT There was, and still is (thank you, Google!), an interface that lets software talk to scanners and imaging devices. It's called a TWAIN. Now, it officially got its name from the famous Kipling line "never the twain shall meet," likely referring to the difficulty in getting software to interface with said imaging devices. However, IT folk, being the cheeky bunch that they are, soon created another meaning for the acronym, which subsequently became folklore throughout the IT industry in the 1990s. This fake, yet more memorable if you ask me, acronym for the TWAIN was "Thing Without An Interesting Name." Thirty years on, and I still remember the acronym.

Why am I telling you all this? Well, here's the thing. I've got an acronym for you! Yes, I know we've already covered one in the book, GROW, but this one is different, trust me. Unless you work in professional coaching, the likelihood of your ever needing to remember GROW is relatively minimal. This acronym I'm about to give you, though, is universal in application, and by that I mean it's not just related to safety or your workplace generally; it's applicable to all the people you will ever meet! A big statement, hey? Have I piqued your curiosity yet?

Since we are talking safety, and particularly safety coaching, I've created a mnemonic to help you remember the acronym. Eagle-eyed readers might have already noticed it in the chapter heading. The acronym I would like you to remember, even if it's the only thing you take away from this book, is B.A.R.

B.A.R. stands for **B**eliefs, **A**ction, **R**esults, and the mnemonic is "raising the BAR on safety." Every organisation has a bar for safety; what's normal, what's tolerated. Now, when you look to influence safety, to raise the bar, I want you to remember this acronym. It will guide you towards raising the bar on safety by drawing your attention to the **B**eliefs of individuals and how they might contribute to the organisation's culture, then to the **A**ctions those beliefs support, and finally to the **R**esults, where we get to see where the bar really is. Aligning B.A.R. with the Integral Model, we can see that **B**eliefs (values, thoughts) align with the inner experience (**UL**), **A**ctions align with behaviours (**UR**), and **R**esults are, well, the results of those **B**eliefs and **A**ctions. Beliefs drive Actions, and Actions give the Results.

Why so much focus on beliefs and values, though? As we've seen through Part I and the Integral Model in this part, people don't just act in response to objective reality; they act based on how they perceive and interpret the objective reality around them. The driving example in Part I again comes to mind as a good demonstration of this. This isn't new science or something I have made up for this book. The sequence I've laid out, **B**eliefs (values, thoughts) acting as a lens that filters reality and drives **A**ctions, which lead to **R**esults, is a well-established set of cognitive-behavioural principles and is considered mainstream psychology. B.A.R. is just a simple way of making these invisible forces visible and having it available for our self-reflection and leadership curiosity.

B.A.R. is predictive. We can logically see how it works to help us understand how **B**eliefs shape **A**ctions, and then what the likely **R**esults will be. The beauty of B.A.R., though, is that it's bidirectional; it can be used in reverse, also! When we already know the **R**esults and preceding **A**ctions, as with an incident, we can then logically work backwards towards the **B**eliefs. For incident investigations, or most any situation where we are trying to understand why a person acted in a certain way, all we have to do is change the sequence from B.A.R. to R.A.B., and we move from a predictive acronym to a retrospective acronym that can be used for sense-making. An example of how this might sound in practice is:

> Given the results and their actions taken (or not taken), how did the person's beliefs or values influence their actions?

We can begin to see how this strongly aligns with New View Safety. R.A.B. guides us to be curious after we know what happened and also prompts us to understand the invisible aspects of safety: the human element. Someone's actions, and the results of those actions, are usually fairly easy for leaders and safety professionals to discover; we have witness statements, image capturing (cameras, video etc.), self-reports, and system diagnostics. What isn't so easy to understand, and what usually gets missed in the Old View of safety, which historically stops at points such as "what rule was broken" and "who is to blame," is the personal beliefs and values of the person involved. For if we know that these beliefs and values are considered mainstream psychology and pivotal to understanding people's behaviours, then surely we must apply this same scientific understanding in the workplace! This alone is one of the main reasons why New View Safety is seen as a more "humanistic" approach to safety, as it uncovers unseen patterns such as beliefs, biases, and values that can influence safety performance. Using R.A.B., we can begin to get curious about individuals' meaning-making and the organisational value signals they might be responding to.

B.A.R. is therefore useful, whichever way you look at it.

With the acronym and mnemonic, I'm proposing B.A.R. (and its retrospective framing of R.A.B.) as a useful first-pass heuristic for leaders who may, through their own self-reflection, realise that a shift is needed from a purely behaviourist, or rule-based approach, to a more systemic and human-centred approach. B.A.R. supports but cuts out some of the complexity of the Integral Model and helps leaders focus on the inner logic behind behaviours. It becomes a quick, memorable tool during conversations, reflections and investigations to guide questions, insights and observations.

As we've seen, B.A.R. can be used as a leadership tool, as well as by individuals themselves as a tool for self-reflection and development. When using it on yourself, it's often easier to begin with the R.A.B. approach as we are often too focused in the moment to stop and be predictive. Whereas looking back on our own actions and results is far easier, and something a lot of people already do. Here are some questions to help start the process for you.

"What results or patterns am I seeing over time?"

"Given those results and actions, what beliefs, assumptions or values might be shaping my behaviour?"

This form of reflection can help leaders to better understand their own belief-to-action patterns and thereby facilitate any professional growth they wish to initiate. This is essentially a part of what professional organisational coaches are doing with their coachees. An important note here in relation to leaders using B.A.R. to understand others' behaviours. B.A.R. is not to be viewed as a prognostic tool; I'm not saying it should be used to label people's beliefs. I'm saying its value comes from knowing beliefs can influence actions, and then, leaders can invite workers to share their beliefs and thereby gain a deeper understanding of motivational factors. Therefore, the leader's job is to create the conditions whereby workers feel safe enough to express their beliefs and values openly, which, as we know from Amy Edmondson, is a critical component in high-performing teams. Think about it this way. A coaching style of leadership is understanding WITH people, not necessarily ABOUT people.

QUESTIONS TO POKE YOUR BRAIN!

- Question 1: In moments of high pressure, how might stepping back and staying curious a moment longer alter the type of information you get from your team?
- Question 2: What might change in your team's resilience if you didn't automatically provide all the answers for them?

6 How New View Safety Benefits from Coaching Style Leadership

HOW NEW VIEW SAFETY BENEFITS FROM COACHING STYLE LEADERSHIP

As I write this book, I am working as a safety leader for a government organisation here in Perth, Western Australia. I can't go into too many details about the journey the organisation has been on over the last 24 months. What I can say is that trust in senior leadership has been devastated, systems have been unfunded and changed without consideration to the point of almost complete ineffectiveness, vast years of accumulated experience have walked out the door, and the reputation of the organisation is hindering every level of talent acquisition needed to operate.

These are not "safety" concerns per se, although psychological safety is of paramount concern for the Regulators and us; they are holistic concerns of the entire organisation. Here's the interesting thing, though. The State Government appointed an interim CEO, brought in to review and rescue the organisation, who isn't just working on systems and processes (**LR**), he's also not overly concerned with people's behaviour (**UR**), as he understands what people have been through and the current systems/processes they are working with. For the last three months, we have been going through a psychosocial safety inventory as a vehicle to help the organisation develop into a "Fearless Organisation" (yes, he's an avid reader and obviously aligns with Amy Edmondson's work).

DOI: 10.1201/9781003590538-8

This involves regular pulse surveys for all workers every six weeks and accompanying team coaching sessions with a trained facilitator from an independent service provider. The framework being used by the service provider and communicated to workers is that of understanding their experiences (**UL**), what they are doing (or not doing) (**UR**), what the current shared meanings within teams and more broadly for the organisation are (**LL**), and how the systems/processes do or don't affect them (**LR**). It's the Integral Model lens that's being applied to this complex situation.

Now, before I go any further, let me comfort those of you who are strategically minded. Systems are being reviewed. Work practices are also being modelled against desired outcomes. Financial plans are being formulated, and appropriate processes are being developed. But this is only part of the picture. The interim CEO knows that you can have adequate systems and processes, but these alone don't guarantee operational performance. Take the current situation I'm involved with. The systems, processes and strategy are all effectively working against even "adequate" organisational performance, and yet the organisation continues to deliver to (albeit degraded and lower than desirable) service levels. How? And more importantly, why?

The goal isn't to just fix the system and double down on rules and compliance; it's much broader than that. It's to increase intrapersonal and interpersonal psychological safety as a mechanism for increasing working relations, communication, creativity, resilience, and ultimately, organisational performance. In Integral terms, we are working on people's inner world and shared space experiences (**UL, LL**) and letting that shape the desirable behaviours (**UR**) and influence the systems (**LR**). All those "fuzzy" or intangible things we can't easily measure or control; that's where the real gold is. By helping people notice and understand their own experiences (**UL, LL**), this process gives people a way to see what is still in their control (after so long without it!), and how they can begin to act locally, not just endure. As the popularly attributed saying to Peter Drucker goes, "culture eats strategy for breakfast."

New View Safety, which defines safety as an emergent property from a complex interconnected system, is a microcosm of the above situation. If we know we need honest, open, and trustworthy relations and communications (read: psychological safety/trust) in order to fully understand and improve the tangible (**UR**), we must take into account all facets of the organisation, including the intangible (**LL, UL**), and how they are all intrinsically linked.

The question then becomes, if New View Safety is what we want, how can we intentionally support behaviour change in organisations without slipping back into blaming and fixing workers?

Try This

Before launching your next safety initiative, map the following:

- "Who will experience what?" (**UL**)
- "What will we ask them to do differently?" (**UR**)
- "How will this fit, or not fit, current norms?" (**LL**)
- "What system changes will support it?" (**LR**)

FACILITATING BEHAVIOURAL CHANGE TOWARDS NEW VIEW SAFETY

One of the biggest challenges a leader will face when beginning to "enact" a New View Safety approach is the entrenched beliefs and behaviours (**UL, UR**) of the people reporting to them. People will behave as they've been trained to up to that date; they too have their own axioms and presuppositions, and depending on how safety has been enacted up to that point, these beliefs (**UL**), which drive behaviour (**UR**), could take a while to change. However, as the leader adopts a coaching leadership style, they will become less directive and more facilitative and curious. As they hold back on quickly solving problems and giving advice to their workers, they begin to ask more questions out of genuine curiosity. This begins the process of empowerment. Slowly maybe. But it will happen. And then workers will begin to take ownership of the "new" direction of safety themselves.

With this approach to leadership, the leader now has a reasoned approach to learning new things about the "work of the organisation," such as learning from normal work, WAI vs WAD, and the contextual and systems influences on actions workers take. With leadership setting the example for others to observe and emulate, and the psychological safety to support it, the ball will start to roll, gathering speed and momentum as it goes. As highlighted by several sources throughout this book, trust is paramount, not just to New View Safety, but organisational performance in general. So, even if you aren't looking to move towards a New View Safety approach, increasing psychological safety (trust) will benefit your organisational performance generally.

Try This

In your next incident review, ask:

"What made this make sense at the time?" before any discussions about "Who is responsible?"

FOSTERING AUTONOMY AND ENGAGEMENT

As highlighted when we covered the definition of a coaching leadership style and the techniques used by MBS, curiosity is a major factor in a coaching leadership style. When workers become less dependent on their leader to provide all the answers, they begin to become more autonomous and engaged in what they are doing and how they are doing it. Research by Deci et al. (2017) in their article "Self-Determination Theory in Work Organizations: The State of a Science" backs this up. It highlights that when people experience increased autonomy (depending on their limits and personal requirements), their motivation, performance, and well-being all improve. This aligns beautifully with the principle from New View Safety of recognising the workers for the experts they are. Frontline autonomy, within clear shared purpose and constraints, allows safe adaptation when reality doesn't match the procedures, or when things unexpectedly change.

Try This

Identify one key decision about safety that you currently make that a competent frontline person could safely make. Hand it over, with clear boundaries, and debrief together afterwards.

OVERCOMING COMPLACENCY AND BIASES

Habituation is a bias whereby workers at the sharp end (for the purpose of this example) stop paying attention to things that happen regularly. The system quirks. Workarounds are needed. Also factored into this is the prevalence of success; most of the time, things go well, so why pay attention to them? With the increased curiosity and facilitative approach of the leader, these system quirks, workarounds, and expectations become exposed and brought to the conscious mind of the workers for consideration. This aligns well with New

View Safety as a way to be continually learning and improving, and building resilience into the systems of work.

Another well-known bias in this arena is the Efficiency-Thoroughness Trade-Off (ETTO). This is where workers can unconsciously prioritise speed (efficiency) over safety (thoroughness) in the quest to meet deadlines, production targets, etc. A leader using a coaching leadership style could be curious about this and ask something like "If you're saying yes to this efficiency, what are you saying no to, in terms of safety?", thereby bringing the bias into discussion and asking for the workers' conscious input into the potential effects, and also the potential solutions. Another avenue for more engagement and autonomy.

In Integral terms, the above biases and your attempts as their leader to surface them are happening across all the Integral quadrants.

UL (workers' inner beliefs and views): beliefs about risk, management and peers, and the presupposition that "this is just how it is."

UR (behaviours): the visible habits that those beliefs drive, such as rushing, skipping steps or checks in a process, and quietly using workarounds.

LL (shared stories): team narratives and "this is how we do it" that normalises ETTO and habituation.

LR (system signals): KPIs, time pressures and other formal signals that reward efficiency and quietly discourage thoroughness.

Try This

In your next field visit, instead of correcting a shortcut immediately, ask, "When you do it this way, what are you assuming about the risk and about what matters most here?" Listen, reflect back on what you heard, and only then explore together what might need to change.

QUESTIONS TO POKE YOUR BRAIN!

- Question 1: When production targets collide with safety protocols, how might your questions unconsciously signal to the workforce which element of the Efficiency-Thoroughness Trade-Off (ETTO) really matters to you at that moment?

- Question 2: What specific signals are you sending that determine whether a worker feels enough psychological safety to share the "bad news" or the "workarounds" that are keeping your systems afloat?

The End

I'm not quite sure how to end this book. So I'll just say thank you for reading it; I truly hope you enjoyed it, or at least didn't hate it. Either way, thank you for taking a chance.

Glossary

Aphantasia: A condition whereby a person cannot voluntarily create mental images in their mind's eye (such as images of loved ones, places, diagrams, or even a representation of self).

Ashby's Law of Requisite Variety: The principle that "only variety can absorb variety," meaning that to effectively respond to the diverse and unpredictable challenges within an environment, a system must possess an equal variety of responses.

B.A.R.: An acronym and leadership heuristic standing for Beliefs, Action, Results. It illustrates how internal beliefs and values filter reality and drive actions, which in turn yield results. Its retrospective counterpart is R.A.B., used for sense-making during investigations to map results and actions back to underlying beliefs.

Coaching Leadership: A behavioural style of leadership that is less directive (command and control) by default, and more facilitative and curious (Socratic and active listening), focusing on support, guidance, and empowerment.

Cognitive Dissonance: The psychological discomfort that results when a person holds two or more competing concepts, beliefs, and/or values.

Confirmation Bias (Attentional Bias): A bias where we believe we already know how an outcome came about, and we seek only the evidence that supports our hypothesis.

Efficiency-Thoroughness Trade-Off (ETTO): A bias where workers unconsciously prioritise speed (efficiency) over safety (thoroughness) in the quest to meet deadlines or production targets.

Empowered Solutioning: A safety improvement approach that utilises aspects of safety knowledge alongside coaching principles, empowering clients to develop their own solutions while leveraging expert knowledge for systemic challenges.

Fundamental Attribution Error (FAE): The phenomenon of attributing blame to others' personality or character flaws in the event of an error, while not treating ourselves the same way.

GROW Model: A foundational coaching framework structured around Goal, Reality, Options, and Will.

Habituation: A bias whereby workers at the sharp end of operations stop paying attention to things that happen regularly, such as system quirks or workarounds.

Hindsight Bias: A bias where we view an incident in reverse, from the end point backwards, and only see a "logical" decision path based on what we already know about how the incident occurred.

Integral Model: A meta-framework (developed by Ken Wilber) used to work with organisational complexity by mapping interconnected dimensions across four quadrants: I (Inner Experience), We (Culture and relationships), Behaviours (What people do), and Systems (Structures and environment).

New View Safety: An evolution of safety practice which utilises social and engineering sciences to account for increasing system complexity. It defines safety as an emergent system property rather than just the absence of injuries, recognising human adaptability as a strength and error as inevitable.

Old View Safety: A traditional perspective focusing almost entirely on preventing accidents, incidents, and injuries by stopping things from going wrong, essentially measuring safety by the absence of negative events.

Organisational Coaching: An evidence-based facilitation approach primarily concerned with unlocking an individual's potential and increasing their performance in the context of an organisation, helping them learn rather than teaching them.

Other-Oriented Perfectionism (O-OP): A bias related to FAE where we unconsciously expect perfection from others at all times.

Psychological Safety: Feeling safe to take interpersonal risks, such as admitting mistakes or sharing bad news without fear of blame or retribution.

References

British Psychological Society [BPS] Division of Coaching Psychology. (2022). https://www.bps.org.uk/member-networks/division-coaching-psychology.

Browning, C. R. (1992). *Ordinary men: Reserve Police Battalion 101 and the final solution in Poland*. HarperCollins.

Bungay Stanier, M. (2016). *The coaching habit: Say less, ask more & change the way you lead forever.* Box of Crayons Press.

Bungay Stanier, M. (2020). *The advice trap: Be humble, stay curious & change the way you lead forever.* Page Two.

Busch, C. (2019). *If you can't measure it.... maybe you shouldn't: Reflections on measuring safety, indicators, and goals*. Lightning Source Inc.

Deci, E. L., Olafsen, A. H., & Ryan, R. M. (2017). Self-determination theory in work organizations: The state of a science. *Annual Review of Organizational Psychology and Organizational Behavior, 4*, 19–43.

Edmondson, A. C. (2018). *The fearless organization: Creating psychological safety in the workplace for learning, innovation, and growth*. John Wiley & Sons.

Entrepreneur's HQ. (2026). *Coaching industry statistics*. https://entrepreneurshq.com/coaching-industry-statistics/

Gallwey, W. T. (1974). *The inner game of tennis*. Random House.

International Coaching Federation [ICF]. (2025). *2025 global coaching study*. https://coachingfederation.org/resources/research/global-coaching-study/

Kunos, I. (2018). Coaching style leadership. *International Journal of Social Science and Business*, *3*(3), 56–62.

Linder-Pelz, S., & Hall, L. M. (2008). Meta-coaching: A methodology grounded in psychological theory. *International Journal of Evidence Based Coaching and Mentoring*, *6*(1), 43–56.

Lloyd, C. (2020). *Next generation safety leadership: From compliance to care*. CRC Press.

NotebookLM. (2025). *AI-generated definition of coaching*.

Passmore, J., & Fillery-Travis, A. (2011). A critical review of executive coaching research: A decade of progress and what's to come. *Coaching: An International Journal of Theory, Research and Practice*, *4*(2), 70–88.

Provan, D., & Rae, D. (2026). *Safety of work* [Audio podcast].

Read, B., & Ritchie, R. (2021). *Safety performance reimagined: A 4D approach to organizational performance*. Amazon Digital Services LLC.

Schein, E. H. (2013). *Humble inquiry: The gentle art of asking instead of telling*. Berrett-Koehler Publishers.

Simons, D. J., & Chabris, C. F. (1999). Gorillas in our midst: Sustained inattentional blindness for dynamic events. *Perception*, *28*(4), 1059–1074.

Wang, Z., Zhang, J., Thomas, M., Yu, J., & Spitzmueller, C. (2023). Coaching leadership and creative performance: A serial mediation model of psychological empowerment and constructive voice behavior. *Frontiers in Psychology, 14*. https://doi.org/10.3389/fpsyg.2023.1077594

Western, S. (2012). *Coaching and mentoring: A critical text*. SAGE Publications.

Index

For Product Safety Concerns and Information please contact our EU representative GPSR@taylorandfrancis.com
Taylor & Francis Verlag GmbH, Kaufingerstraße 24, 80331 München, Germany

www.ingramcontent.com/pod-product-compliance
Lightning Source LLC
LaVergne TN
LVHW010915011026
849280LV00025B/275
* 9 7 8 1 0 3 2 9 6 7 1 9 6 *